Mining and Regional Development

DEREK SPOONER

Theory and Practice in Geography

OXFORD UNIVERSITY PRESS · 1981

Oxford University Press, Walton Street, Oxford OX2 7DP

London Glasgow New York Toronto
Delhi Bombay Calcutta Madras Karachi
Kuala Lumpur Singapore Hong Kong Tokyo
Nairobi Dar es Salaam Cape Town
Melbourne Wellington

and associate companies in
Beirut Berlin Ibadan Mexico City

Published in the United States by
Oxford University Press, New York

© D. J. Spooner 1981

All rights reserved. No part of this publication may be reproduced,
stored in a retrieval system, or transmitted, in any form or by any means,
electronic, mechanical, photocopying, recording, or otherwise, without
the prior permission of Oxford University Press

This book is sold subject to the condition that it shall not,
by way of trade or otherwise, be lent, re-sold, hired out, or
otherwise circulated without the publisher's prior consent,
in any form of binding or cover other than that in which it is
published and without a similar condition including this
condition being imposed on the subsequent purchaser

British Library Cataloguing in Publication Data
  Spooner, Derek
  Mining and regional development. — (Theory and practice in geography).
  1. Mining industry — Great Britain
  2. Great Britain — Economic conditions
  3. Great Britain — Regional planning
  I. Title II. Series
  338.2'0941 80–41597
ISBN 0–19–874045–X

Set by Oxprint,
and Printed in Great Britain
by J. W. Arrowsmith Ltd, Bristol

Contents

1. INTRODUCTION
2. MINERAL RESOURCES AND MINING LOCATION
3. 'EARLY' STAGE DEVELOPMENT—THE 'MINERALS-DOMINANT' PHASE
4. 'MIDDLE' STAGE DEVELOPMENT—MINING REGIONS AS DEPRESSED AREAS
5. 'LATE' STAGE DEVELOPMENT—THE MINING REVIVAL AND THE RETURN OF THE RESOURCE FRONTIER

REFERENCES

Preface

The renaissance in mineral exploitation in the United Kingdom during the last decade has aroused very considerable interest in potential benefits to the national economy and in environmental hazards, but much less in the consequences for regional development. Awareness of this neglect provided the initial stimulus for this volume. Examination of the contemporary role of mining in regional development in Britain seemed to lead naturally to an analysis of its role in the past, and to comparative studies of experience in other countries. The approach adopted therefore has a historical perspective, and while the final chapter deals largely with the United Kingdom, earlier chapters draw upon examples from the Third World, the United States, and to a lesser extent, Western Europe. No attempt is made to include the non-capitalist world, which would clearly merit another volume.

The approach to minerals is deliberately a catholic one; there are no intended exclusions. If fuel minerals dominate much of the discussion, this reflects their enormous importance in regional development patterns.

Many of the themes presented I have first put forward in an unpublished paper at a British-German seminar on regional development organized by the Institute of British Geographers. I am grateful to the University of Hull for financing my attendance at that seminar, and to John House, a fellow-participant, for his encouragement to develop the paper into a more extended form. Other debts are owed to John North for his collaboration in studies of British coalfields, and to Graham Robertson, who introduced me to some of the more obscure literature of mining geography. The figures were drawn by Keith Scurr. The final manuscript owes much to the editorial guidance of John House and to my wife Christine's critical scrutiny. The faults remain my own.

Skidby DEREK SPOONER
October 1979

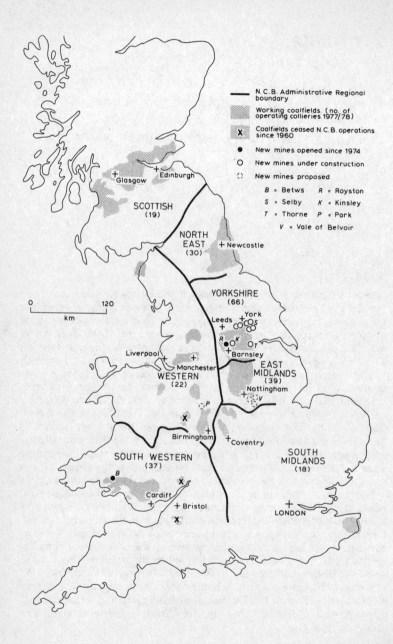

Frontispiece. UK coalfields in 1979.

1 Introduction

Mineral exploitation and the 'new geography'

During the 1960s, when human geography became thoroughly infused with the nomothetic approach, and spatial or locational analysis became the primary obsession, the exploitation of mineral resources received comparatively little attention as a subject for geographical inquiry. There were exceptions. Goodridge (1967) was examining the historical geography of Cornish metal mining; House and Knight (1967) were studying community implications of colliery closure in County Durham; Odell (1963) and Manners (1964a) were involved in aspects of the economic geography of world energy supply. But the apparent simplicity of the locational problem in the extractive industries led most devotees of the 'new geography' to concentrate their attention upon the seemingly more complex problems posed by location of manufacturing and retailing, by land-use patterns in cities, by the spacing and size of settlements and by the development of transport networks. The lack of interest in mining geography was nowhere more apparent than in the spate of books attempting to synthesize the key concepts of the revitalized discipline. For example, Abler, Adams and Gould (1971) devoted only two pages (of nearly 600) of their 'geographer's view of the world' to mineral resources.

However, although spatial analysis occupied the limelight, other new strands in the subject were developing, to which the study of mining was potentially more important. Two in particular can be identified—'regional development' and 'resource management', heirs to 'regional' and 'applied' geography respectively. Both have flourished in the 1970s.

Regional development and resource management

The growth of these fields of study reflects changes in geographical methodology, but also mirrors the times in which we live. The philosophy of traditional regional geography has been undermined by the growth of the nomothetic approach, but also by the effects of the Industrial Revolution, which dissolved the rural, local, largely self-sufficient way of life (Wrigley, 1965). 'Regional geography' has therefore been supplanted by 'regional development', a field of study where the emphasis is placed upon the evolution of regional economic problems, and upon regional policy and planning, a

2 Mining and regional development

change further stimulated by the emergence of 'the regional problem' as a persistent feature of the advanced economies.

The growth of resource management as a field of geographical endeavour can be linked to the long-standing central position in geography of man–land relationships, and to the revival of the subject's 'ecological' tradition. But the strongest impetus has come from the onset of the 'environmental revolution'. This transformation in man's attitudes to resources and environment manifests itself in debates about pollution, limits to growth, soft energy paths (Lovins, 1977), and the 'small is beautiful' philosophy (Schumacher, 1973). In the 1970s, the oil crises have intensified the interest in resource depletion.

Mineral exploitation has important linkages with these geographical fields and has a poor image in relation to both. In regional development studies, the main significance of mining is usually seen as the creation of regional problems—as in South Wales, Appalachia, or Wallonia. Regions are stranded by mining's demise. Policies to aid such regions have traditionally emphasized the switch of labour and capital into other sectors, particularly manufacturing. Preoccupation with such problems has tended to make us forget that there have been phases in history when mining has played a positive role in regional development, giving rise to new patterns. In the 1970s there are a few signs, however, that the purely negative view of mineral exploitation is being modified, as policy-makers have sought to broaden their armoury of weapons to attack intractable problems. The prospect of 'windfall profits' for mineral-producing areas arising from the escalating price of some minerals, especially oil, is also leading to some reappraisal.

If the mining industries, especially coal, have had a poor image in regional development terms, this has been doubly true in the context of environmental impact and resource management. George Orwell's image of the mining area, observed on the road to Wigan Pier, persists—'the monstrous scenery of slag-heaps, chimneys, piled scrap-iron, foul canals, paths of cindery mud' (Orwell, 1937). The bitter remarks of Harry Caudill (1962), in his study of the Cumberland mountains of Kentucky, that 'coal has always cursed the land in which it lies . . . it is an extractive industry which takes away all and restores nothing', captures a common sentiment. Mining companies are frequently regarded as environmental vandals. Old images die hard.

The search for new mineral sources in many parts of the world, stimulated by a variety of market and political factors, increasingly brings conflict between economic and environmental objectives. The examples of Alaskan oil, Vale of Belvoir coal, and Orkney uranium reflect the same resource management dilemmas. Discussion of the contribution of such mineral exploitation to regional development cannot be divorced from a consideration of the environmental problems posed. It is equally true, but often ignored, that development of these economic activities should not be debated only in environmental terms.

The intention of this volume is to review the relatively neglected role of mining activity in regional development, examining the interaction between mining and regional problems and policies. The framework adopted is time-based, examining the role during different stages of the process of economic growth. As the foregoing pararaph stresses, in the late stages of development, reached in the advanced nations of Western Europe and North America, interaction with environmental problems becomes intense. In this sense, the economic geography of mining can be seen as representing increasingly an area of convergence between regional development and resource management issues.

Regional development: a framework

Regional development studies focus upon spatial aspects of development, the latter being viewed broadly as the whole process of change brought about by the creation and expansion of an interdependent world economic system (Brookfield, 1975). Development is not simply economic *growth*; it has both positive and negative aspects (growth, decline, stagnation). Indeed in the context of mineral exploitation, decline is a primary concern. Nor is development exclusively measured by economic criteria such as GNP *per capita*; criteria of social welfare and quality of life are increasingly important. Development is fundamentally about inequality, and this is a spatial as well as social, racial or occupational problem. Spatial inequalities can be examined at a variety of scales; regional development focuses upon inequalities within nations at an intermediate scale.

A basic concept is cumulative growth. Once a region (for whatever reason) gains an initial advantage over others, various forces tend to widen rather than reduce the gap between them; once triggered, growth proceeds by 'circular and cumulative causation', while spatial interaction with other regions tends to heighten the regional disparities by the process of 'backwash' (Myrdal, 1957). Although eventually counteracting 'spread' effects may develop, these tend to be weaker, and cannot prevent the development of a 'centre-periphery' or 'heartland-hinterland' structure. The hinterlands or peripheries of a nation develop a dependent relationship to

4 Mining and regional development

the industrial heartland or centre, supplying it with resource inputs, and lagging behind it in development. Increasingly such relationships are depicted as forms of internal colonialism. When economic growth is sustained over long periods, some equilibrating tendency may occur and progressive integration of the space economy may result, through outward flows of growth impulses down the urban hierarchy and catalytic impacts on surrounding areas (Berry, 1973). However, the power of the 'centre' to continuously innovate new dynamic forms of activity (Thompson, 1968) is just as likely to leave the periphery condemned to a lower rate of growth.

Although many writers thus have little faith in the long-run equilibrating effects of regional growth processes, it does appear that the principal types of regional problem to emerge, and (less convincingly) the extent of regional disparities may be related to the stage of national development attained. This is well illustrated by Friedmann (1966), who suggests a variety of models on which the following synthesis is based. Spatial development is related to stages in the evolution of the national economy.

1. In *pre-industrial societies*, economies are highly localized; there is no national, integrated spatial framework.
2. In *transitional societies*, characterized by rapid industrialization and 'take-off', dramatic spatial shifts occur, especially in association with movement from agriculture to industry. A dualistic centre-periphery structure develops as investment is concentrated in a favoured area. The periphery begins to lag, though resource exploitation may trigger growth in some areas. Regional development levels diverge strongly.
3. In *industrial societies*, approaching economic maturity, considerable problems of regional backwardness emerge. Some areas are stranded by depletion of their resources or a drop in demand for their major product. They need conversion to a new economic base. Regional policy becomes prominent. In time strong peripheral sub-centres *may* develop and promote regional convergence.
4. Finally, in *post-industrial societies*, a space economy with a functionally interdependent system of cities develops. More speculatively, the periphery is absorbed and regional balance achieved. Metropolitan organization, urban renewal and environmental quality become dominant planning concerns.

In relation to this model, regions are classified according to the problems that they pose for development. *Core regions*, consisting of clustered metropolitan areas, are characterized by high promise for economic growth. *Upward transitional areas* are those where natural endowments and location relative to the core region suggest the possibility of greatly intensified use of resources (through spread effects). *Resource frontiers* are zones of new settlement developing virgin territory. Of particular importance in the context of this study is the non-contiguous resource frontier, an isolated pocket of development frequently based upon large-scale investment in mineral resources. Resource frontiers may involve substantial urbanization, with the city as an agent for transformation of wilderness. Problems arise in integration with other regions and in achieving permanence. *Downward transitional areas* are long-established settled regions, whose particular resource combinations suggest less intensive development may be appropriate. They include declining mining regions. Such regions present major planning challenges. Finally, *special problem regions* are a diverse group requiring specialized policy approaches, such as international frontier or tourist zones. The whole set of development regions form a spatial system based upon the core region.

The plan of this book follows the above framework. After a chapter describing the crucial features of mining industry, three chapters examine mining's role in the 'early', 'middle' and 'late' stages of development, corresponding broadly to stages 2, 3 and 4 in the evolution of the national economy. Each stage is characterized by a dominant theme—respectively the resource frontier, the depressed area and the mining revival.

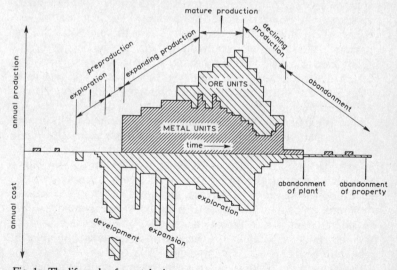

Fig. 1a The life cycle of a metal mine
Figs. 1a and 1b from W. C. Peters, *Exploration and Mining Geology* (John Wiley & Sons, Inc., 1978). Reprinted by permission.

2 Mineral resources and mining location

Non-renewable resources and mining cycles

Natural resources are commonly divided into two major groups, for which a variety of terms are used, indicating the character of the basic dichotomy. Mineral resources can be equated with 'stock', 'inorganic' or 'non-renewable' resources, by contrast to the biological resources, which are 'flow', 'organic' or 'renewable'.

Probably the most vital characteristic of mineral resources in regional development terms is that, at least on the human time scale, they cannot be renewed. They are exhaustible, fugitive and finite. They can be used once and then are gone (though it is true that we have made some progress in recycling, as with steel scrap, which may extend their usable life). To the resource manager, the fugitive nature of mineral resources poses fundamental dilemmas—*when* shall development take place, at what rate shall the resource be depleted?

Minerals are subject to the discovery-depletion cycle (Fig. 1a), which may be seen as having six stages—discovery, development, expansion, mature production, decline and exhaustion—experienced by individual mines and by whole mining districts. The passage of the cycle may be extremely erratic; mines may be rejuvenated or even resurrected, usually as a result of technological or market change. Peters (1978) illustrates this process of rejuvenation in the life cycle of the San Francisco (Oatman) gold-mining district of Arizona (Fig. 1b).

The length of the normal life cycle is highly variable. Some mines have short colourful histories, giving quick bonanzas, but cannot sustain long-term deep operations; others survive for decades despite disappointing returns. It is ignored too often that many mines (and mining regions) persist successfully for long periods, despite lowering grade of deposit, increased depth and tightening economic conditions. The famous Almaden mercury-mine in Spain has been in continuous operation since 1499; the varied ore district of Santa Eulaila, Chihuahua, Mexico, now approaching exhaustion, has been mined seriously since 1702 (Dixon, 1979). Comparisons with the product-life of manufactured goods or longevity of manufacturing plants need not necessarily be invidious. Sant (1972) showed, for example, that over 10 per cent of all inter-urban

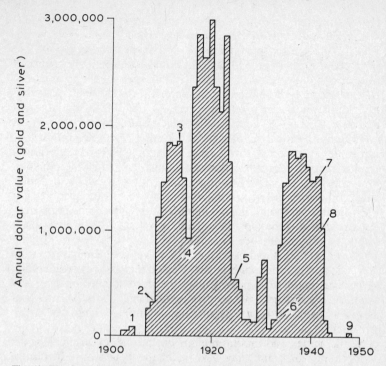

Fig. 1b The San Francisco (Oatman) mining district, Arizona 1. Early production from exposed ore bodies. 2. Major mine began production from small ore bodies. 3. New company formed to develop deeper ore. 4. Major mine began production from deeper ore bodies. 5. Major mine closed. 6. Price of gold raised from $20.67 to $35.00 per ounce, followed by redefinition of reserves, and reopening of mines. 7. War-time shortages of labour and equipment. 8. War-time closing of gold-mines. 9. Post-war mining attempt.

manufacturing moves in Britain ceased production within 5 years; manufacturing industries face constant problems of technological and market change, while the demand for some mineral products remains remarkably constant over long periods.

The cyclical nature of mineral exploitation is critical in regional development because of the varying intensity of investment and employment requirements at different stages. Generalizations are difficult because of the variety of minerals and methods of extraction. Those types of deep mining which utilize labour underground make much higher labour demands during the mature phase of producing than those, like oil wells, which do not, and where labour demand is much heavier in the development phase. Surface mining presents a different case again. Particularly in resource frontier regions, mining operations may require heavy infrastructural investment, especially of the economic overhead capital type such as water supply, transport facilities and power sources, but also possibly in social overhead capital—education, health, housing and other community services. In such circumstances, the total labour requirements for exploitation of a mineral deposit may peak well before the maximum production phase is reached. A modern, and extreme, example is described by Mackay and Mackay (1975). In examining the direct impact on Scotland of North Sea oil exploitation they distinguish four major overlapping phases: exploration, manufacturing (i.e. fabrication of production facilities), production and construction. Their 'construction' phase parallels the other three, and includes such temporary activities as pipe-laying. For each field, the second phase—manufacturing plus some temporary construction—is the most labour-intensive and lasts three to seven years; employment levels fall in the third phase, which may last up to 30 years and is the most capital-intensive. Problems of adjusting to cyclical labour demands in the affected regions are considerable. A contrasting case is provided by the development of the Selby coalfield (Yorkshire), where labour demands upon the total economy will remain relatively small-scale until the labour-intensive production phase.

While mines and mining regions go through cycles in their development, the same may be true of the mineral production of nations. Hewett (1929) postulated a five-stage cycle. The first stage culminates in the maximum export of crude ore; in the second the maximum number of mines is in operation; in the third the maximum number of smelters is in operation; the fourth sees the maximum production of metal from domestic ores; the final phase brings reliance on foreign ores as the main source of supply for smelting industries. Hewett's model was developed at a time when states like the UK that had reached the ultimate phase could still control the resources of states or colonies in stages one and two. However, Robertson (1974) points out that in the 1970s the importing states are much less able to control the mineral resources of the exporting nations (OPEC provides eloquent testimony), and argues that a sixth stage has ensued for countries like the UK, which are now seeking to augment foreign supplies by turning again to indigenous sources, galvanized by rising prices and increased political uncertainties.

The supply of minerals and resource evaluation

Mineral supply can be regarded as relatively elastic compared to biological resources, where the responsiveness of supply to price changes is limited in the short run by the length of natural growth processes, the problems of land-use conversion and the law of diminishing returns. Wrigley (1962) considered that the great limitation on industrial growth before the Industrial Revolution was the reliance upon organic fuel sources (timber, charcoal). The shift to coal was crucial as the scale and rate of industrial growth was no longer regulated by agricultural productivity or biological growth rates. Given adequate technology and capital, fuel supply could be relatively easily increased. (Today, however, the large scale of enterprise and capital investment has reduced the short-term responsiveness of supply to price; a modern underground coal-mine may take ten years to attain full production.)

The switch to minerals as the basis for industrial growth brought fundamental changes to the spatial pattern of economic activity—an emphatic shift to a sporadic or punctiform distribution (Wrigley, 1962). This had important implications for infrastructural development, especially transport. The heavy traffic flows generated by localized mining could sustain investment in transport facilities much more easily than the diffused areal production associated with exploitation of biological resources. The change to mineral resources also has profound implications for regional development, leading swiftly to the burgeoning of new patterns of economic activity and infrastructural development.

Evaluation of the potential supply of minerals available for exploitation in any region as a basis for development is a complex matter, particularly as the perception of any resource does not rely only on physical properties, but on a range of 'cultural' factors. The term 'resource' does not apply to a material or an object, but to a value placed upon a material, in view of the function it may perform or the operation in which it may take part (Blunden, 1977). 'Resources are not, they become' (Zimmerman, 1951)—they expand and contract in response to human wants and deeds and to technological, economic and political conditions. Thus, minerals have *time value* (Peters, 1978). Some minerals (like flint) become ex-resources, reverting to 'neutral stuff', made obsolete by technological change.

In resource taxonomy the term *reserves* is usually limited to those mineral deposits which can be exploited under existing economic conditions, and with available *technology*. Possibly we should add under existing *political* conditions. They are classified in a variety of ways (e.g. Manners, 1969), using such terms as measured, proved, inferred, possible, etc., while the broader term 'resources' also includes other mineral deposits known but not economically or technologically recoverable, or inferred to exist but not yet discovered. 'Reserves' are thus developed from 'resources' through the application of technology, capital and expertise and in response to cost and price change.

The imprecision of the resource concept is such that the business of estimating reserves is a confused one, with wide margins of error and inconsistencies of practice. Reserve estimates tend to be highly dynamic, and must always be evaluated in relation to the assumptions used in their calculation. Odell and Rosing (1976) demonstrate that the difference between *technically* and *economically* recoverable reserves is dependent upon investment decisions.

Mining location

If mineral resources have a time value it is also true that they have a *place value*, and this is especially important in understanding the role of different minerals in regional development.

The comparative lack of attention to mining in location theory reflects a belief that mining represents 'the simplest case of the locational problem' (Abler, Adams and Gould, 1971). Coal has to be mined in Castleford and not in Cambridge. However, while it is obvious that minerals cannot be extracted where they are absent, it must be recognized that the multiplicity of deposits of variable size, quality and location presents the entrepreneur with a considerable problem of locational choice, and one made more complex by the possibilities of substitution between alternative mineral products. Two variables of basic importance in deciding which deposits are exploited are the quality of the deposits and the location of the market.

McCarty and Lindberg (1966) hypothesize that there is a relationship between the value to weight ratio of a mineral and the distance from the market of the location at which exploitation will occur. Value to weight ratio, or unit value, is crucially affected by the extent of opportunities for product substitution, which influences the elasticity of demand for a particular mineral product. Thus products like sand and gravel command a low unit value. Such minerals are easily substitutable and have high demand elasticity; their market price is strongly affected by distance and transport costs. They are said to have high *place* values—their value is highly dependent upon the place in which they are located. Unit value and place value tend to be inversely related, but other factors such as transportability must also be considered—liquid products which can be easily conveyed by pipeline or sea tend to be exploited further from markets than unit values might suggest.

Hay (1976) introduces greater sophistication to the mining location theory. He demonstrates the variable viability of a set of mines with variable production costs (an important factor) located at different distances from the market. He makes the fundamental, but previously neglected, point that price (unit value) is not an independent variable, but will itself be affected by the interplay of demand and supply.

Wilson (1968) discusses the question of the location of individual mines within a mining field. He argues that the least-cost principle traditionally stressed as the basis for manufacturing location can be utilized for pit location, with the mining company seeking to minimize its combined production and transportation costs. Some empirical evidence appears to support this argument. The recent choice of site for the NCB drift mine near Selby has been strongly influenced by transport and mining costs. The site is adjacent to existing rail routes (minimizing investment in new transport facilities), while at the same time avoids the water-bearing strata of the Bunter sandstone, through which shaft sinking would have been extremely costly (Forrest, 1976).

Nevertheless, the same difficulties that have beset theories of manufacturing location with regard to the influence of non-economic factors—political, behavioural, etc.—also limit the power of these theoretical models to explain real-world mining distributions. Optimal patterns are rarely achieved. Warren (1973) summarizes the geography of mining as 'the response of enterprise to the known geological facts, though conditioned by the availability of capital and labour, by political consideration, planning controls, and very importantly, by the legacy from past patterns'. The location of individual mines is constrained increasingly by social, political and environmental considerations. Thus while site choice by the NCB at Selby described above does suggest a least-cost principle, the same organization already claims that they have rejected the optimum sites for their proposed Vale of Belvoir mines on environmental grounds.

Mineral classifications

The discussion so far has been largely couched in broad generalizations about mining activities that conceal a wide variety of products, techniques and modes of operation. Traditional classifications of minerals by economic geographers have often used inconsistent criteria. For example, Jones and Darkenwald (1965) classified partly on the basis of physical/chemical characteristics and partly on the basis of end-usage.

12 Mining and regional development

Blunden (1975) also uses more than one criterion, but his classification is useful in the current context because it introduces a vital distinction of a locational nature, and draws upon the concept of place value. His five categories are: (i) ubiquitous non-metalliferous minerals, (ii) localized non-metalliferous minerals, (iii) non-ferrous metals, (iv) ferrous metals, (v) carbon and hydrocarbon fuels—the sources of heat and power. Blunden further subdivides the *non-metalliferous* minerals into five groups based upon unit value, but seeking to emphasize the interrelationships between unit value, frequency of deposits, degree of processing, scale of working and market radii.

The *ubiquitous* non-metals groups are largely used in the construction industry (including sand and gravel, limestone, igneous and metamorphic rocks, brick clays); they have low unit values and high place values; they are frequently interchangeable and substitutable; they tend to be exploited near markets. In the UK many of those used for building aggregates double in price within 50 km of the source. Traditionally such minerals are worked in a dispersed pattern and individual workings are small to medium in scale. The workings tend to be located in relation to existing patterns of population concentration. They are region-serving, rather than region-forming. They tend to contribute to the agglomeration of economic activity around existing centres, rather than initiating new patterns.

By contrast, the *localized* non-metals have much higher unit values. They include most of the raw materials for the chemicals and fertilizer industries like salt and potash, as well as china clay and fluorspar, minerals which are highly restricted in occurrence, are often produced on a relatively large scale and can stand greater transport costs to more distant markets. In regional development terms, such minerals may play some part in region-forming, initiating new patterns and attracting processing industries. This is possibly true of the Cheshire salt deposits. In other cases, however, mining remains relatively isolated and the minerals are withdrawn to supply industry in other regions, where industrial growth is already well entrenched and external economies of scale are powerfully developed. The Cornish china clay industry is an example. Although the Hensbarrow area has developed into one of the world's major producing districts, all the china clay leaves the region. No china clay consumers have been attracted to Cornwall—major customers like paper-making and pottery industries find greater locational advantages elsewhere.

Apart from the distinction between localized and ubiquitous minerals, the other major features of significance in the foregoing classification is the identification of the mineral fuels. These differ from the other groups in that they are not incorporated in the final product (petro-chemicals are an obvious exception), but are rather consumed to provide energy for direct industrial and domestic consumption, as well as for electricity generation. In regional development, the significance of these fuels has been immense. In the early years of steam power the huge quantities of coal used at low levels of efficiency, plus the relatively high cost of coal transport, drew industry to the coalfields.

Time, scale and technology

Classifications which include locational criteria apply at a fixed point in time; the place value of mineral deposits can change through the influence of such variables as technological change and economies of scale. In most mining industries, these variables interact to produce a trend towards larger-scale mines or quarries, often serving larger markets over greater distances. Technological change in transport plays a vital permissive role—the development, for example, of liner trains, motorways, larger trucks, giant ore-carriers, pipelines and conveyors all tend to allow production at greater distance from the market, often working lower grade deposits.

Thus in some formerly 'ubiquitous' mining industries in the UK, economies of scale in both production and transport have led to increasing localization in certain areas. In the 1960s and 1970s, low value minerals used in building aggregates have begun to be transported over much greater distances in response to growth of demand and exhaustion of more accessible deposits. The higher costs of widening market radii have been partly contained by improved means of transport and by further on-site processing (Blunden, 1975).

The last twenty years have seen a world-wide trend towards massive mining operations, using large equipment to reap scale economies. Both mines and processing facilities have become heavily capital-intensive; labour has become a diminishing factor, though often still important because of rising wage rates. Bosson and Varon (1977) point out, for example, that while US copper production has doubled over the past 30 years, ore handled has increased fourfold because of decreasing grades, but related manpower has fallen by 60–70 per cent.

In many mining industries, large open-pit mining and improved processing technology have reduced operating costs considerably, making possible the economic exploitation of low grade deposits. In the USA, the average value of metallic ore mined in 1969–70 was about $5 a ton for open-pit mining, compared to about $12 a ton for underground mining, with roughly the same difference for non-metal ores. The cutoff grade for many ores has decreased. In the early 1960s, a cutoff grade of about 0.8 per cent copper was considered the economic minimum for porphyry copper deposits—by 1970 it had reduced to 0.4 per cent, despite the fact that copper prices had only increased 20 per cent in real terms. The opening of the Wheal Jane tin-mine in Cornwall, in 1971, reflected the development of new technology to handle a highly disseminated ore body not amenable to economic treatment by the old Cornish beneficiation methods.

The world industry today is dominated by open-pit rather than underground mining. Bosson and Varon (1977) calculate that in 1970 some 65–70 per cent of all ore mined in the world came from open pits, while in the US the proportion was as high as 90–95 per cent. *The Mining Annual Review*'s survey of Western world production of 22 major minerals (excluding coal and oil) shows that the number of very large mines is growing steadily. In 1968 there were 155 such mines each producing 3 m. tonnes p.a. or more; by 1978 212, of which 157 were open-pit. In total, this survey recorded over 7,000 mines, but the 1079 producing over 150,000 tonnes p.a. accounted for over 90 per cent of world output. The emphasis on open-pit mining is far greater if one considers the lower value non-metals. For example, more than 95 per cent of non-metal production in the USA comes from surface operations. Low unit value minerals can rarely stand the higher costs associated with underground mining.

A corollary of increased scale of operation is concentration of ownership and increased control of production by large corporations, many of which are involved in all stages of the mining cycle, as well as being vertically integrated with mineral-using manufacturing. According to the same MAR survey, the bulk of Western world mining is dominated by about 160 companies. Multinational operations are the norm. In some industries like oil, the oligopolistic character of production is pronounced.

As a factor in regional development patterns, the emphasis on increasing scale of operation is of considerable importance. Regional concentration and specialization of production is encouraged and might be seen as a movement towards region-forming rather than region-serving activity. The ability of the industry to initiate new patterns may be enhanced; the impact of individual new mines is potentially great. However, such arguments must be treated cautiously. Much mining is highly capital-intensive and makes very limited impact on regional employment. Furthermore, the organizational characteristics of the industry remain highly centralized. The existence of centre-periphery structures may mean that large-scale exploitation of new mineral resources acts largely to sustain the growth of the centre by providing resource inputs.

One sure effect of increased scale of operations, however, is to increase the potential for conflict with environmental considerations, recalling the theme of convergence between regional development and resource management issues.

3 'Early' stage development—the 'minerals-dominant' phase

Mineral resources and national development in the Third World

A brutally simple description of the end-product of Bolivian tin mining is provided by Lindqvist (1972):

> The road between Potosí and Oruro crosses a mining landscape in ruins. Abandoned smelting furnaces, rusty conveyor belts, the relics of the exploitation of centuries. Everywhere the traces of former wealth. Left is simply a land of extreme poverty. It is only along the road between Potosí and Oruro that children stand with outstretched hands screaming for bread. Someone had been there, taken what he wanted, and decamped.

Here the path from resource frontier to depressed area appears to have been a direct one.

Such an emotional picture does not make it easy to assess the role of the mining industry in either regional or national development. However, it does suggest strikingly why in recent years the mining activities of international corporations in Third World countries have come under increasingly critical scrutiny. Conflicts have arisen between the corporations and host nations over the sharing of benefits from mineral development; resentment of dependency, neo-colonialism and external control has led sometimes to violent results.

Nevertheless, historical evidence does suggest that the role of a nation's mineral resource base in fostering economic growth is greatest during the early stages of the process. The dangers of using historical analogy to argue the potential importance of mineral resources to today's developing economies are great, but some experts do believe that mineral wealth may lubricate the growth path, especially where balance-of-payments problems are acute. The discovery and shipment of oil from Ecuador in the early 1970s brought an immediate improvement in the country's balance of payments, as well as providing revenues from concession rights that helped the government reduce its recurrent budgetary deficit (Gilbert, 1974). For such reasons mineral wealth is eagerly sought in many developing countries. Spengler (1961) argues that resource shortage could have a growth-restricting effect, while resource abundance could be growth facilitating. Advantages of a productive mineral resource base may include stimulus for new techniques, ability to attract foreign capital that would not otherwise be forthcoming, and opportunities to develop exports. McDivitt and Jeffery (1976) point to benefits in three categories—finance, infrastructure and 'environment for development', but Grunwald (1964) in an earlier, more balanced discussion, pointed out that many of these benefits are reduced by the weakness of 'spillover' effects from resource exploitation. Profits are leaked to foreign capital, the infrastructure is often specialized and outward-orientated, linkages with other activities are limited, few of the resources produced are used as inputs for domestic industry. Morse (1964), for example, argued that between 1900 and 1950 the Venezuelan government received in taxes and royalties only about 10 per cent of the value of exports as payment for depletion of the country's oil wealth, and income accruing to the Venezuelan factors of production used was negligible.

Initiating new regional patterns: American experience

Arguments about the *national* aspects of economic growth are beyond the scope of this volume, but we can see that in most developed countries a growing independence of economic activity from specific localized resource bases has appeared, as the ends of economic activity become increasingly non-material, and as the quantity of natural resource inputs declines relative to total inputs (Adler, 1961). In consequence, it is likely that it will be in the early phases of development that the positive influence of mining activity on *regional* patterns will be most pronounced, with the exploitation of minerals triggering growth in new resource frontier regions.

Few comprehensive studies examine the role of natural resources in the regional development process, but one notable exception is provided for the USA by Perloff and Wingo (1961). They delineate a number of periods in American development. After the 'early agricultural period', based on productive land accessible to the eastern seaboard, came the 'minerals-dominant' phase. From *c.*1840 the expansion of railroads and manufacturing stimulated demand for mining products. In the period between 1870 and 1920, 'developmentally dominant effects emerged from the growth of the minerals economy, shifting rapidly among regions, triggering, intensifying or transforming the nature of regional growth patterns'.

The importance of minerals lay not only in their direct contribution but also in the nature of their linkages with succeeding stages of production. This applied especially to coal and iron ore, which, because of heavy transport costs, attracted the iron and steel industry to areas like West Pennsylvania. It applied to a lesser degree to the development of oil and natural gas as a trigger to growth in the South West. Oil production produced fewer regional linkages and was more capital-intensive, though the additional presence of sulphur and salt accelerated the expansion of chemicals industries. It applied least to the mountain region of the West, where metal ores production did not induce any substantial expansion of linked activity.

The 'minerals-dominant' period was succeeded by 'deepening' of the American economy, with regional resource effects diminishing. The power of the 'market magnet' became the major influence, as resource inputs fell as a proportion of total inputs, and as the service sector blossomed. Resources remained important, however, through their influence on the *inherited* patterns of population and industrial distribution. Finally, from mid-century an 'amenity resource' era set in, with shifts in population and economic activity towards Florida, the South West and the Pacific coast.

Perloff and Wingo's analysis draws upon the export base theory (North, 1955), whereby regional growth is promoted by the ability of a region to produce goods or services demanded by the national economy, and to export them at a competitive advantage with respect to other regions. Thus the metals of the mountain states and South Western oil were regional 'exports' triggering development. A flow of income into the region is produced, which through the multiplier effect tends to expand regional markets for goods and services.

In time it can be argued that the external determinants of regional growth become less important, that internal structural change (an increase in the proportion of employment in manufacturing and services) and expansion of local markets lead to self-reinforcing regional growth, that internal factors become important in determining regional growth rates, including external economies, and that the internal organization of production becomes more important than the dominance of export production.

The export base theory of economic growth has attracted considerable criticism both on theoretical grounds and in terms of its empirical validity (Keeble, 1967), and sophisticated analyses suggest that other variables like wage levels are also important regional growth determinants. The export base theory is a much simplified explanation of regional growth and may be most appropriate to its early stages, when mineral exploitation is often important.

Mining and regional growth: the mechanisms

Review of Perloff and Wingo's study has suggested some of the ways in which the impact of mineral exploitation and 'export' are felt by regions, and these need to be elaborated further. The rise in income, production and employment in mining stimulates the expansion of other activities through the *multiplier* process.

Labour moves in to work the mines, and services develop to meet the consumer demand thus created. Local food production may be stimulated. The Californian gold rush of 1849 led to the opening up of large crop and livestock areas, which eventually became independent of the mining industry (Warren, 1973). In the mining region of northern Chile agricultural exploitation of non-irrigated areas has historically been closely linked with the mining camps, but here as the veins of metal were worked out, the agricultural villages declined too (Blakemore, 1971). The demand for mining equipment may lead to the development of manufacturing. Thus the Cornish metal mining industries of the last century fostered a network of ancillary industries including explosives, drill and safety fuse manufacture. Such industry generates further rounds of job creation and local expenditure on goods and services.

Forward linkages may develop within the region. With metallic ores, the first stage may be some form of beneficiation, such as concentration, pelletization or sintering; the second stage may be smelting. Such developments greatly increase the value of the product. For example, in the US iron ore industry over 90 per cent of all ore produced is now pelletized or sintered—this more than doubles the value per ton of ore produced (McDivitt and Manners, 1974). The third stage may be the attraction of fabricating plants. Foward linkage to mineral-using industries is of prime importance in the regional development process.

Particularly in virgin areas, infrastructure will be developed. Out of necessity, rather than philanthropy, much of this may be provided by the mineral developer. Expenditure on infrastructure may exceed that on mine construction. Thus of the initial expense of $100m. on the Lamco mining project in Liberia, capable of producing 7.5m. tons of iron ore per year, 50 per cent was on harbour and railway construction and 10 per cent on electricity supply, a township, hospitals, etc. Only 32 per cent was spent on the prospecting, planning and construction of the mine (McDivitt and Manners, 1974). The company town is a characteristic feature; in many remote locations, the dearth of speculative builders, or of local governments capable of providing housing, forces companies to set up their own towns.

Truncation of the process

The multiplier effects may set in train cumulative and self-reinforcing growth. However, it is vital to stress that this is not automatic. Much of the impact may leak from the region, or is minimized by the nature of modern mining activity. Baldwin (1966) provided a good example when he described the failure of the Zambian metal mining industries to generate 'spread' effects since 1920. The mining industry was capital-intensive, no local industries had been set up to utilize the mines' output, nor did the mines provide any significant number of Africans with incomes sufficient to transform consumption patterns—effective demand was insufficient, for example, to provide more than a limited stimulus to food production. In the advanced economies, an excellent example is provided by Appalachia. This is a region where natural resource exploitation in the past (including a non-mineral, timber) did not lead to a prosperous regional economy. Instead, modern America's major depressed area has been created. A diagnosis of the region's ills was provided by the President's Appalachian Regional Commission (1964):

Where a society depends primarily on the extraction of natural resources for its income and employment—as did the people of Appalachia—it is extremely important that a high proportion of wealth created by expansion be reinvested locally in other activities. The relatively low proportion of native capital did not produce such a reinvestment in large sections of the region. Much of the wealth produced by coal and timber was seldom seen locally. It went downstream with the great hardwood logs; it rode out on rails with the coal cars; it was mailed between distant cities as royalty checks from nonresident operators to holding companies who had bought rights to the land for 50 cents or a dollar an acre. Even the wages of local miners returned to faraway stockholders via company houses and company stores.

Thus the benefits of mining development in a resource frontier may be leaked abroad or to a 'heartland' elsewhere in a variety of ways. The life span of intensive development may be short, and, because of the low labour requirements of much mineral exploitation, creation of urban agglomeration may not occur (Richardson, 1976). The skilled workforce may be largely imported 'gringos', who stay for short periods and transmit much of their income from the region. Backward and forward linkages may form with industries that are not located in the same region, or even the same country. Only about 30 per cent of the minerals produced in the developing countries are processed there, and this proportion has remained constant since 1950 (Bosson and Varon, 1977). The mining development may remain an enclave with comparatively little effect on its surrounding region.

The flow of profits from the region is a 'backwash' effect that often enhances the growth of the centre from which development is controlled. The wealth created is reinvested elsewhere, and the centre-periphery relationship may be intensified. Two examples of the way in which resource frontier development at the periphery may stimulate the expansion of the centre have been described by Friedmann (1973) and Odell (1973) for Chile and Venezuela respectively. Both countries are highly dependent on mineral exports.

In Chile, the development of nitrate and copper-mining enclaves in its northern region has had a powerful effect on the national economy through tax revenues from mining corporations. The considerable government revenues so derived resulted in heavy expenditure on public works with special emphasis on the capital city, Santiago, in central Chile. Here an urban upper class, grown wealthy in mining, banking and commerce, adopted a sumptuous life-style based on conspicious consumption of imported goods. According to Friedmann (1973), hyper-urbanization was stimulated; Santiago became a highly primate city in the Chilean urban hierarchy. Hyper-urbanization retarded economic development, diverting scarce resources from investment in highly productive activities, reducing the propensity to save and discouraging agricultural production. The development of resources at the periphery was a catalyst to excessive concentration at the centre.

In Venezuela, the Maracaibo oilfields became isolated enclaves of modern technology amid areas of mainly subsistence agriculture in the 1940s and 50s. The oil was piped direct to ports, with little impact on intervening areas. The 'gringo' oil communities aimed to achieve a high level of self-sufficiency in service provision. Initially there was a considerable multiplier effect on Maracaibo City, which became a centre for the co-ordination of oil company activities, but improvements in air transport and telecommunications led to a shift to national headquarters in the Venezuelan capital, Caracas, at the expense of the regional centre. Moreover, government oil revenues were expended in a highly skewed spatial pattern, favouring the Caracas region, much of them mis-spent on non-essential building projects (Robinson, 1971). By the early 1960s one-fifth of Venezuelan population was clustered in the Caracas metropolitan area, and the centre-periphery structure had been perpetuated.

The foregoing discussion raises many questions. Are some minerals more likely than others to lead eventually to cumulative growth? Which minerals in the past played an important role? Can they do so today? What factors determine whether the mining region is converted from a 'tributary' to a 'comprehensive' style of development? Are there 'good' and 'bad' minerals from this standpoint? For further illumination of these questions it may be instructive to examine the role of coal and other minerals in nineteenth-century Europe.

22 Mining and regional development

Coalfields in nineteenth-century Europe

Coalfields were the growth regions of nineteenth-century Europe, though not all grew at equal speed. Wrigley (1961) demonstrated the close linkage between coal output, industrial growth and population change in the *early* phase of industrial development between 1850 and 1914, in the belt of coalfields stretching from French Pas de Calais through Belgium to the Ruhr. With coal transport expensive and labour and capital more mobile, he found that the variable quality and cost of extraction of coal in the different subdivisions of the zone affected directly the pace of industrial growth, outweighing national influences. The growth of population was intimately linked to the growth of coal output.

In describing the overhauling of Saxony by the Ruhr between 1882 and 1907, Tipton (1976) indicated that coal had a dual role. The northern Ruhr was being opened up along the Lippe river: 'the huge new mines drew population into the Ruhr and increased export income. The indirect effect of the mines in encouraging other industries within the Ruhr was even more important.' Metal manufacture was sensitive to the location of coal; the mining and heavy metal industries grew together.

Similarly in Britain, Brown (1972) argued that coal not only had 'an original and dominating role as a source of energy', but also 'was a big enough activity in itself for its distribution to have direct economic significance'. Its direct effect on regional income, output and employment was considerable, and it drew a huge labour force to the coalfields, while at the same time it exercised a powerful locational pull on other industries. This last was especially true before cheap transport developed, and while efficiency of coal use remained low. In Weberian terms it made little sense to transport large quantities of coal to non-coalfield locations.

In 1900 well over half the British towns with a population of 50,000 and over were situated on or near coalfields. The coastal fields were especially important, benefiting from access to tidewater for export. Major staple industries developed, like steel and shipbuilding, though Hall (1973) stresses that 'each coalfield also sustained a host of varied crafts and trades'. However, some features of coalfield development presaged twentieth-century problems. First, the outward orientation of infrastructure in the coastal fields, with emphasis on transport links to ports for overseas and coastwise shipment, was to prove a disadvantage once emphasis on the home market for goods became of key importance (Caesar, 1964). Secondly, development of the coalfields did in some ways contribute to the national centre-periphery structure. In 1800 London, with a population of 900,000, was already easily the largest city in Britain. The growth of London played a vital role in the Industrial Revolution. According to Wrigley (1969) it had much to do with coal-mining expansion; in the eighteenth century about one-sixth of the whole British output was shipped down the east coast to London. The effect was reciprocal: London stimulated coal expansion, access to coal for domestic and industrial consumption helped to sustain London's growth. The centre-periphery relationship was already developing.

In Britain, no other mineral equalled coal in its impact on nineteenth-century regional growth. Iron was important, but initially this was often exploited in the coalfields from blackband ores. Subsequently the Jurassics of Cleveland were prominent, aiding the strong regional growth of north-east England from mid-century. In the twentieth century ore location has been a weaker influence on the location of new iron and steel plants, but the heritage of nineteenth-century patterns has remained of key importance in an era of awakened social conscience and full employment policies (Warren, 1976).

The impact of other metals was also considerable locally (tin and copper in Cornwall, lead in the Pennines), producing rapid population growth and sometimes important infrastructural development, albeit of a specialized nature. But the speculative nature of the mining economy in such areas, with violent variations in output, did not encourage stable long-term development.

Mining and regional growth: the crucial factors

The experience of nineteenth-century Europe does suggest the importance of minerals in giving rise to new patterns, and above all illustrates the special role of coal. It is clear, however, that the extraordinary importance of coal related to conditions unique to that period of technology—a period when the wastefulness of coal use dictated a large scale of production with a huge labour force, and drew industry to the coalfields.

If we turn therefore to countries now undergoing the 'early' stages of economic development, it is clear that the historical experience is unlikely to be repeated, even should they possess substantial coal reserves. Technology today is vastly more economical in coal use. The world has shifted to a multi-fuel economy, and oil, a highly transportable fuel, has replaced coal as the major energy source. Perloff and Wingo (1961) showed that even in the relatively early phase of the oil economy the direct localized linkages with other economic activities were more limited than had been the case with coal, while the industry was also much

24 Mining and regional development

more capital-intensive. Brown and Burrows (1977) are in no doubt about oil—'the tendency to attract industry to the places where it is extracted . . . has been slight'. Odell (1963) was forced to conclude that the development of the USA's huge oil-based industrial complex on the Gulf Coast was probably unique, though there were prospects of parallel developments elsewhere. The limited extent to which oilfields have developed mature industrial regions was attributed partially to the structure of the oil industry. The international major companies had no need or incentive to increase the oil consumption on the oilfields by encouraging industrial development. Indeed, because of government pressures in the developing countries, they have been forced sometimes to sell their oil more cheaply in the area of production than to the rest of the world—a powerful incentive not to encourage local oil offtake.

Thus fuel minerals today are not likely to exercise the same propulsive effects on regional development that coal produced in the past. The lessons of coal lie rather in the way its impact was achieved.

The foregoing discussion suggests that certain characteristics of mineral exploitation are important if the transition is to be achieved from an 'export'-orientated 'hinterland' region to one of self-sustaining growth. Here we are stressing characteristics of the mining industry and mineral products; lest this approach appears unduly deterministic, it must be added that social and institutional characteristics of the region concerned may be equally important. The effectiveness of the export base in stimulating regional growth may depend considerably upon the ability of the region to organize itself for economic growth, and the size of the multiplier effect may depend upon community capacity for social development. Having made this reservation we can postulate four main characteristics that will influence mining's ability to sustain regional economic expansion.

First, both the supply and demand for the products of the region must be an enduring one. The mining industry must be able to support over a long period an extensive stream of nationally or internationally demanded products, in competition with other regions. Fields which rapidly exhaust their richer deposits, or cannot stay cost-competitive, are unlikely to be much use in long-term regional development.

Secondly, large scale is crucial, preferably with a large labour force. This was the advantage of coal, but we also see it occasionally with other minerals. It was a factor in the growth of the Rand goldfield to the biggest industrial agglomeration in Africa. Gold was discovered in 1886 and Johannesburg had grown from a mining camp to a city with a population of 150,000 by 1904 (Warren, 1973). Growing material needs were supplied locally, coal was found at near-by Witbank to supply fuel—the scale of the whole mining industry became sufficient to lead the region (and the South African nation) into sustained growth.

Thirdly, the mining industry, should generate localized backward and forward linkages. It should have *complementarity* with other activities (Richardson, 1976). Here lay the strength of nineteenth-century coal: it was able to pull user industries to the coalfields.

Finally, a 'good' mineral resource development must be characterized by little 'leakage'. A substantial proportion of the returns from mineral 'exports' must find their way into the region, and stimulate demand for regionally produced goods and services. Additionally the region must become attractive to migrants.

Resource frontiers in regional policy

Many developing countries in the 'early' stages of development already face severe problems of spatial disparity between a fast-growing metropolitan centre and a laggard periphery. The problem is often complicated by the presence of huge empty areas, as in Brazil, to which national aspirations may be directed. Regional policies may therefore develop at an early stage. In the development of the periphery, despite the numerous problems already discussed, mineral resource frontiers may be regarded as regions of new opportunity around which such policies may be structured.

In particular, mineral development may occasionally be seen as a possible basis for the development of a new growth pole. The theory of growth poles has become confused or even debased (Moseley, 1973, Brookfield, 1975) and a variety of terms (pole, point, centre) are used, but a growth pole policy can be viewed basically as an attempt to simulate in a backward region the conditions favourable for growth. Such policies may be employed variously to regenerate depressed areas, to combat the growth of metropolitan regions, or to institute growth in underdeveloped regions. Boudeville (1966) stressed the importance of a set of expanding industries generating a *propulsive* growth effect from an urban area on a surrounding zone. Of particular importance are the ability of such industries to form complementary relationships with other economic activities (forward and backward linkages), the multiplier effects arising from expenditure on wages, etc., and the psychological impact on the regional economic environment. In time the desired effect should be a restructuring of the region and a cumulative and rapid increase in regional output. The choice of appropriate instruments as a basis for such a policy is extremely difficult, especially in underdeveloped 'empty' regions with few comparative advantages, where the exploitation of mineral deposits may seem virtually the only possibility.

Once again, Latin America provides the best documented examples. The comprehensive survey by Stohr (1975) of 75 regional development programmes operating in that area in the late 1960s revealed 19 settlement and resource frontier programmes, of which 3 were based on minerals. These were for 'Norte' Chile, 'Comahue' Argentina and 'Guayana' Venezuela. Of these the last is undoubtedly the most dramatic.

The Venezuelan project is attempting to use mineral wealth as a new focus for the development of its remote and empty eastern region, while at the same time promoting a counterweight to Caracas. It is interesting that the finance for this scheme stems largely from oil revenue, which previously had stimulated central development. 'Sowing the oil' was a key slogan of the Bettancourt government, which created the Guayana Corporation in 1960.

The new industrial complex is based on a variety of minerals among which rich iron deposits are outstanding, and on hydroelectric power, and is centred on a new city. 'Ciudad Guayana', founded in 1961, had reached a population of 130,000 by 1975. According to Friedmann (1973) the experiment could already be counted as a success; Ciudad Guayana had become Venezuela's principal centre for heavy industry, including steel, and an important exporter of intermediate products. Forward linkages had thus developed in the way demanded by growth pole theory. Nor should the symbolic role be underestimated. Some reservations remain, however. According to Stohr (1975) the multiplier effects had been modest, while for all its comprehensive planning the project had been Caracas-controlled. Moreover, we may legitimately ask what are the opportunity costs for Venezuela of this massive development.

Clearly, the problems of utilizing mineral wealth as the basis for a regional growth pole are considerable; mining is often capital-intensive and integration with the existing regional economy may be hard to achieve. There is a risk that a 'cathedral in the desert' will be created. The major lesson of the Venezuelan example may be the value of a comprehensive approach to planning; the mechanisms are too delicate to be left to chance.

4 'Middle' stage development— mining regions as depressed areas

This chapter examines the development of problem mining regions in the more advanced economies, drawing principally upon British and American experience.

Mining and change in the economic structure

In the USA the 'minerals-dominant' phase was followed by one in which the market became the major locational attraction and resource effects greatly diminished, as structural shift from agriculture and mining to manufacturing and services accelerated. Spengler (1967) estimated that whereas less than forty years ago nearly 30 per cent of the labour force needed to be located close to the natural resources, only 7 per cent were still resource-bound. By 1970 mining's share of direct employment in the USA had fallen below 1 per cent, and the contribution to national income had fallen to just over 1 per cent.

It could be argued that mining remained more important for regional growth than its share of national income and employment suggested, because it was uneven in its spatial distribution and provided the start for many production sequences. However, a much greater range of minerals was in demand, many of which could be used as a substitute for, or in combination with, others, thereby diminishing the locational attraction of individual sources, while many minerals had become more transportable through the development of on-site processing techniques.

Similar changes have occurred in the UK. By 1973 the mining and quarrying industry employed only 1.4 per cent of the national workforce, and the contribution to GNP had fallen to below 2 per cent, though since 1973 under the influence of North Sea developments the proportionate contribution has risen slightly—to 2.2 per cent of GNP in 1976.

The changing structure of mineral production

Important changes have also occurred in the structure of mineral production. In the USA *fuel* production since 1940 had grown at the average rate for all minerals, but other *non-metals* had expanded much faster, while *metal* production remained static, affected by economies in use, substitution and increasing imports (Estall, 1972). By the late 1960s, despite low unit value, the total value of the output of non-metals had become twice that of metals.

The growth in importance of the non-metals is significant in the regional development context. Estall (1972) shows that, unlike the fuels and metals, significant production is found in every state. The largest producer, California, is the most populous state. Half the total output by value is provided by eight states, which account for 48 per cent of the US population. There is a broad correspondence with the geographical distribution of population. As suggested in Chapter 2, the 'ubiquitous' non-metals (especially the constructional minerals) tend to be region-serving, rather than region-forming. This has the side-effect of presenting major physical planning problems; demand for these minerals is greatest in zones of intense land-use competition, and where concern for environmental quality is often great.

While employment in the relatively ubiquitous non-metals sector has grown, that in the localized fuels and metals sectors has fallen, with serious regional implications. Depressed mining regions have emerged. A good example is the Upper Great Lakes iron-ore region, mainly in Minnesota and Michigan, still the major producing area in the USA. Here, job losses in a relatively isolated physically difficult area have created a considerable regional problem (Estall, 1972).

Post-war experience in the UK is comparable (Table 1).

TABLE 1
Production of selected minerals in the UK

		1956	000s tonnes 1966	1976
Fuel:	Coal	228,500	178,900	123,800
	Natural gas* (including colliery methane)	27	260	57,269
	Crude oil	67	78	11,630
Metals:	Iron ore and stone	16,506	13,877	4,852
	Tin	1.7	2.1	4.0
Non-metals:	Sand and gravel	64,300	108,300	117,681
	Limestone	32,803	67,850	86,034
	Igneous rock	14,562	28,713	37,215
	Sandstone	4,514	7,668	13,522
	Fluorspar	76	149	217

*Approximate coal equivalent.
Source: UK Mineral Statistics, 1977.

In the 1950s and 1960s the volume of output of the constructional non-metals grew rapidly, while the leading fuel (coal), and metal (iron) experienced serious decline. Between 1972 and 1974 the tonnage of sand and gravel produced for the first time exceeded that of coal. The differences in labour intensity remained vast; even after a decade of catastrophic job losses, the coal industry still accounted in 1973 for more than 86 per cent of all employment in mining and quarrying (Table 2, GB only).

TABLE 2
Employment and output in GB mining industries

	Employment (000)		Gross output (£)	
	1963	1973	1963	1973
Coal	593.3	304.4	860.9	793.0
Oil and natural gas	—*	2.2	—*	145.0
Metals	6.3	—†	14.2	—†
Stone and slate	21.1	19.0	70.6	229.6
Chalk, clay, sand and gravel	21.5	21.4	76.9	231.1
Miscellaneous	6.3	3.7	27.9	40.4
Total	648.5	350.7	1050.5	1439.3

*1963 figures for oil and natural gas included in 'miscellaneous'.
†1973 figures for metals not available, iron ore mining being classifed under the iron and steel industry. Other metals in 'miscellaneous'.
Source: UK Mineral Statistics, 1975, 1977.

In 1973 payments by the coal industry of wages and salaries accounted for £496.2m. out of £584.9m. paid by all mining industry, though in terms of capital expenditure coal had been overtaken by the rapidly growing oil industry. Out of £365m. total capital expenditure in 1973 by mining industries, £102.3m. was by coal and £210m. by oil (UK Mineral Statistics, 1977).

The decline of coal

In both the UK and USA employment in coal has dwarfed other minerals, and the rapid contraction of employment in this mining industry has given it immense significance in regional change. The decline in employment can be attributed to several factors: saleable output has fallen because of competition in fuel markets from other sources of energy (oil, natural gas, nuclear power) and because of economies adopted in key industries like steel-making; employment in the industry has fallen faster still as new methods of coal-mining and handling have been applied. Changes in labour productivity have been rapid; between 1953 and 1963 US coal output fell by 0.4 per cent p.a., but employment by 8 per cent p.a. In the 1960s US output began to grow again by 2.6 per cent p.a. but employment continued to fall by 1.2 per cent p.a.

Regional problems are intensified by the uneven pattern of decline arising from the coalfields' variable geological endowment, length of working and market access. In the UK there has been a long-established trend towards concentration of production in the 'central' coalfields of Yorkshire and the East and South Midlands. During the rapid contraction of coal production since 1960, rates of decline have been much higher in the longer-worked 'peripheral' fields. In the North-East, for example, the number of operating collieries fell by 80 per cent between 1960 and 1977, output by 62 per cent, employment by 70 per cent. Some small fields have ceased production entirely, the most recent, Shropshire, in 1979. In the 'central' coalfields output fell by less than 30 per cent, employment by around 45 per cent. As a result of these changes, the 'central' fields increased their share of saleable output from 48 per cent to 62 per cent, of employment from 39 per cent to 53 per cent. Coal production has become increasingly concentrated in the belt of country running north from Coventry almost to York (Fig. 2) (North and Spooner, 1978b).

The declining industrial attraction of the coalfields

While the sheer scale of contraction of the coal industry has had grave consequences for many regions, another factor has been of at least equal importance. The switch of many consumers from direct use of coal to the use of electricity has been of major significance in regional development.

When coal was used in bulk for steam-raising for industry, the pull to coalfield locations was strong, and the initial absence of adequate means of long-distance transmission for coal-generated electricity enhanced the attraction of coalfields. However, as transmission facilities developed, with the first National Grid constructed in Britain between 1932 and 1936, the pull of coal weakened. Electric power became increasingly available over wide zones at uniform cost, permitting decentralization of manufacturing industry and giving much greater freedom of locational choice. Electricity pricing policies, in the UK at least, give no advantage to the consumer who locates close to the power station, except in the case of a few bulk users, like smelting for aluminium.

Coalfields remain attractive to the location of power stations, though other bases for electricity generation now compete. In both the USA and UK over 50 per cent of coal output goes to the electricity generating industry. But the rash of power stations constructed since the war in the UK's 'central' coalfields do not provide any significant local or regional attraction to industrial consumers.

The potential advantages of the coalfields in providing cheap power have been negated by the development of the grid, pricing policies, and the construction of nuclear and oil-fired power stations in coal-deficient areas. For much manufacturing industry a localized *primary* fuel source, coal, has been replaced by a ubiquitous *secondary* source, electricity.

In a much larger country the above argument must be qualified; greater spatial variations in electricity costs may occur. Even in the USA, however, as Smith (1971) showed, only a few small areas have significantly high or low costs, and in most of the country electricity costs are relatively uniform.

Obstacles to coalfield 'conversion'

As we saw in Chapter 3, many coalfields appeared in the nineteenth century to possess the critical qualities necessary to develop self-reinforcing growth. In the twentieth century, however, the success of many such areas has proved illusory, and many have slipped into a *downward* spiral of cumulative causation. The very scale of past mining activity has ultimately proved a burden, as a huge and specialized workforce has been decanted on to the labour market at colossal speed. Between 1955 and 1968 approximately one million coal-mining jobs were lost in the major Western producing countries (Brown and Burrows, 1977). There were two obvious primary outcomes for many coalfields (although experience varied greatly): heavy male unemployment and considerable out-migration.

Why have so many coalfield regions failed to sustain their growth as coal has declined? Two reasons have already been suggested—the huge scale and rapidity of job losses, and the diminished locational attraction to coal consumers. Nevertheless one must seek additional reasons for the failure to attract enough new activities to utilize the considerable labour surplus, which included not only unemployed miners but also a large under-used female work force. What have been the obstacles to 'conversion' of coalfield economies?

A major problem has often been the low level of labour demand throughout the national economy. This was the particular problem of UK coalfields in the early 1930s. By contrast, the buoyant labour demand in West Germany in the 1960s clearly assisted the conversion of the Ruhr.

Secondly, the decline of mining in many coalfields coincided with the decline of other industries that were linked in location. Many of the great industries like textiles and shipbuilding that had flourished in many European coalfields were badly affected by the new world economic order that arose after 1918, while changing technology and resource requirements have rendered many coalfield iron and steel sites obsolete. A region like north-east England has thus faced the problem of adjusting to the simultaneous decline of several major industries. Decay in these has had a cumulative effect upon a range of ancillary industries as well as directly reducing labour demand and regional spending power.

Additionally, some characteristics of the coalfields may have proved inimical to the establishment of new industry. The concentration of effort in nineteenth-century British fields in a few massive staple industries represented an 'overcommitment' of resources, which inhibited the growth of newer forms of production (Richardson, 1965). In some ways this argument appears at odds with Hall's conclusion about the diversity of industry on the coalfields in 1900, but it is true that many coalfield industries, especially engineering, became highly specialized to serve the needs of the staple trades like shipbuilding and textiles. It is the exceptional experience of the coal-based Birmingham-Black Country conurbation that may be instructive. This area was isolated from the sea and never developed the same export orientation and 'heavy' engineering industry as the coastal fields. Its host of small-scale metal trades proved a suitable basis for the evolution of light engineering industries in the late nineteenth century, from which in turn evolved the mass production of electrical goods and interchangeable components for cycle and motor car manufacture (Lee, 1971). It was these modern growth industries that signally failed to develop on the coastal coalfields.

The over-commitment to massive industries like coal and steel had other implications. Many coalfields lacked enterprise as a result of their particular industrial history. The oligopolistic structure of their industries concentrated decision-making into the hands of a few. Regions like South Wales were thus denied communities with wide traditions of enterprise, and potential entrepreneurs turned instead to chapel and to schoolroom (Manners, 1964b). In other cases, domination by outside companies denied opportunities to local entrepreneurs to gain industrial experience. This was true in the Southern Illinois coalfield, 1900–40. Brown and Webb (1941) showed that local entrepreneurs were excluded from direct participation in the coal industry and instead turned to real estate promotion, sparking a speculative boom that in time intensified the problems created by the sudden demise of mining. Clearly, however, experience of individual fields has been diverse.

Although evidence is limited, it may also be true that the industrial structure of coalfield regions has influenced the availability of capital. According to Chinitz (1960), regions with traditions of small-scale and highly competititive industries tend to be more ready sources of capital than those with heavy industrial traditions. The coalfields did not develop institutions and traditions assisting pioneer entrepreneurs to obtain capital.

A further factor inhibiting the development of new industries in British coalfields was the relative poverty of external economies in these areas compared to the developing industrial agglomerations elsewhere. This relates again to the oligopolistic structure of major coalfield industries; the characteristically large firms provided many of their own services, which were not therefore available to the industrial community at large. Moreover, the specialized nature of the industries concerned—coal and steel especially—meant that many of the services (like coal research laboratories) were unsuitable for the use of other industry (Manners, 1964b).

Furthermore, industrial conversion in depressed mining areas may be hindered by social and economic 'rigidities', which preclude the adaptation of existing resources to the needs of expanding sectors. The labour force may not only be specialized but also conservative and opposed to change, while union strength may deter potential investors. The fixed capital of the mining industry is also inflexible; it is not easy to find new uses for coal tipples and winding gear. Derelict mining land can be reclaimed for a variety of purposes, from agriculture and forestry to industrial estates, but such treatment is costly.

Thus the legacy of mining has often been inimical to the development of other activities. Spoliation of the environment has become increasingly significant, as there is now considerable evidence that regional environmental quality is a potent influence on the location of manufacturing plants. In small countries like the UK many manufacturing industries can now locate freely over wide zones within their margin of profitability, and precise choice of location may be made on the basis of criteria such as residential desirability. Studies of firms on the move have confirmed that detailed economic evaluation of alternative locations is rare and that much decision-making is influenced by personal impressions and images of an area (e.g. Townroe, 1971, Green, 1974). Keeble's analysis of industrial change in the UK supported the hypothesis that 'the image of particular localities as attractive residential environments is now in the 1970s a factor of major importance in manufacturing growth, strongly influencing the locational decisions of both workers and industrialists' (Keeble, 1976).

It is important to stress that we are dealing with the perception of regions, rather than objectively described regions. Many coalfields have a poor *image*. They are part of a 'legendary North'. The survey of management in the Northern Planning Region by House *et alia* (1968) showed that this legendary north did exist in the minds of managers and their wives—one of 'regional isolation, an unpleasant climate and most of all with an unfavourable urban, and industrial landscape'. Negative images of 'northern' coalfields in the UK have undoubtedly hindered the attraction of new firms from southern England.

The preceding discussion presents a sharp contrast to the relatively optimistic picture of coal's impact on regional development drawn in Chapter 3. The advantages of large-scale and locational pull argued there have not proved unassailable, and in time have rebounded upon many coalfields as the combined effect of collapse in coal and other industries has triggered a downward spiral of decline. At the same time the specialized and 'rigid' nature of development in some coalfields, allied more recently to an unfavourable residential image, has thwarted the growth of alternative activities. The scale and persistence of the regional problems created have forced governments to seek remedies.

Regional policies

Reviewing the development of regional policy for depressed mining areas is made difficult by the fact that policies have rarely been formulated for such areas alone. With this reservation, we can identify four main types of policy, before looking in detail at British and American experience. Policies may be operated by government institutions with varying powers and responsibilities working at several different levels—international, national, regional and local.

(a) *Industry protection policies*. *Sensu stricto*, these are not regional policies, but they can have significant regional effects. National governments may intervene to support an industry for a variety of reasons, including softening the effects of decline in stricken regions. Several examples are found in the European coal industries. In the UK, markets for coal have been protected by tax on fuel oil and by government influence on the Central Electricity Generating Board's (CEGB) policy, a striking recent example of which was the pressure on the CEGB to order the Drax 'B' coal-fired station in 1977 (North and Spooner, 1977). The ECSC (European Coal and Steel Community) has smoothed the downward transition of many coalfields, for example by protecting high-cost Belgian mines by levies on lower-cost German and Dutch.

(b) *Labour transference policies*. Moving workers to work is often advocated as a regional policy measure, but rarely practised. The European coal industry provides something of an exception; the ECSC has provided financial assistance to mineworkers in a variety of forms to improve their occupational and spatial mobility. The NCB's Inter-Divisional Transfer Scheme, started in 1962, was described by Moyes (1974) as 'probably the biggest peacetime exercise in the planned movement of manpower in the UK'. Mineworkers made redundant by colliery closure were transferred in large numbers both intra- and inter-regionally.

In the UK the earliest attempt to attack the problems of the depressed areas was the 1928 Industrial Transference Board, which provided finance to cover the cost of workers' removal to the new areas and the setting up of training centres. Some 250,000 people moved in this way between 1929 and 1938, but the movement of labour hardly constituted a solution to the depressed area problem at a time when there was a labour surplus throughout the country. Since 1945 support for labour mobility has been relatively neglected.

(c) *Policies of attracting new employment.* The major constituent of regional policy in many European countries since 1945 has been the attraction of new industries to the depressed areas. Varying sets of 'assisted' areas have been defined, in which mining regions have been prominent, with grants, loans or tax concessions available to private firms for capital investment in new plant, buildings or machinery. In some cases labour subsidies or premiums have also been available (as in the UK between 1967 and 1976), while control of development may be exercised in 'non-assisted' areas by a certificate or permit system. Factories may be built for firms, either to a standard 'advance' pattern, or to meet individual requirements. A variety of ancillary measures all work to the same end—the conversion of the industrial base to new, growing forms of industry. More rarely, state-owned industry may take the lead in creating new jobs. The effect of these policies is not necessarily that the new industry immediately employs ex-miners, but tends to be more diffuse, increasing the availability of new jobs in the local labour market *as a whole* (Bulmer, 1978).

(d) *Infrastructural policies.* Alternative or complementary to the direct encouragement of new industry is the indirect approach of infrastructural investment, aiming to improve the environment in which industry can operate. Such investment may give direct support for productive investment in the form of economic overhead capital (roads, harbours, water supplies, etc.), or else may aim at improving human resources, through investment in such social overhead capital as education and health facilities. In view of the recognized role of good environment as a stimulus to industrial growth, the potential importance of this type of regional policy in the resuscitation of mining areas can hardly be overstressed.

The spatial pattern of investment that is followed has been a matter of considerable controversy in regional policy, and one of direct relevance in mining regions where existing development is scattered. Most spatial policies initially focused upon areas of need—spreading aid to localities where problems were most intense, as indicated by unemployment rates. In time most countries have questioned the wisdom and cost of such policies of 'grapeshot regionalism', which apparently accept existing settlement patterns as permanent. The advantages of concentrated development have been increasingly championed. Growth centre policies have been adopted at a variety of scales, attempting to simulate in problem regions the conditions favourable for economic expansion, primarily through the concentration of infrastructual investment.

UK regional policies and mining areas

In few countries have the problems of depressed coalfields dominated regional policy to the extent found in the UK. The first major legislation designating assisted areas, in 1934, created four Special Areas, all coalfield based; the North-East, central Scotland, South Wales and West Cumberland. Although since 1945 the assisted areas have been extended to include many peripheral rural regions (notably in the Development Areas created in 1966), two changes have provided continued emphasis on mining areas.

First, the rapid rundown of coal-mining in the late 1960s forced the government to introduce Special Development Areas in 1967, a classic 'worst-first' policy, selecting localities within the Development Areas for preferential rates of aid. Most SDAs were in the coalfields. Coupled with this policy, the major series of advance factory building programmes since 1962 have stressed coalfield locations, especially since 1967 when a 'rolling programme' was introduced for the SDAs whereby advance factories were replaced as soon as tenants had been found.

Secondly, the report of the Hunt Committee (1969) on the problem of areas neglected by previous legislation concluded that the most intense problems to be found outside the DAs were in the Yorkshire coalfield, arising from contraction of manpower requirements in mining. The initial list of Intermediate Areas established in 1969 included both the Yorkshire coalfield and the Erewash valley in Derbyshire, an area with similar problems.

The Hunt report was significant in other ways. It emphasized the need to improve the poor environment in mining regions. It stressed the importance of infrastructure to the growth environment and the need to invest in carefully selected 'points of opportunity'. Reclamation of derelict land was regarded as a high priority. Derelict land clearance grants were fixed at 75 per cent in IAs compared to 85 per cent in the DAs; in 1975 these rates were raised to 100 per cent.

The foregoing account has stressed the close relationship between the depressed coalfields and regional policy. The demise of other mining industries has also contributed to regional problems and the drawing of the assisted areas map. The disappearance of Cleveland and Furness iron and Cornish copper, the rundown of Pennine lead, Cumberland iron and Cornish tin, have all left pockets of distress. All these areas have had assisted area status at various times. Indeed in the most recent and drastic revisions of regional policy in 1979, Redruth, formerly a great centre for Cornish mining, has been promoted (demoted?) to SDA status.

While regional policy has continued to be dominated by schemes to lure industry by direct cash aid to depressed areas, policies to reshape the infrastructure have developed more strongly at the intra-regional scale. White Papers on central Scotland and north-east England in the early 1960s advocated growth zone approaches. In the North-East, concentration of investment has been actively practised in County Durham.

Here central locations for industry have been encouraged on the major communications routes, to provide employment for people travelling or relocating from mining areas. To the east, the New Town of Peterlee had been designated in 1947 to provide centralized development for a scatter of mining villages. The logical corollary to identification of growth points within regions (unless the population is growing fast) may be identification of places for decline, though because of political and social sensitivities, this is rarely spelled out. Durham County Council have been exceptionally bold in this respect. Their development plan of 1951 classified 357 settlements into four categories, of which the 114 'D' villages were to be gradually run down as the population moved away. The contraction and eastward shift of mining was leaving many villages shorn of their original *raison d'être*. The possibility of allowing indiscriminate development throughout the unplanned and unco-ordinated settlement pattern was rejected, as this would spread limited financial resources too thinly and lead to only a marginal improvement in the county's economic and environmental circumstances (Blowers, 1972).

Inevitably the policy has attracted opposition and has undergone some revisions. The basic strategy remains unchanged, however, and a few villages have now completely disappeared. Critics argue that increased personal mobility can now support a pattern of more dispersed settlement and that many of the condemned villages still contain valuable communities, close-knit by ties of kinship, class and common origins. The policies are said to be preoccupied with 'work' and 'place', at the expense of 'folk' (Bulmer, 1978). Others, however, have concluded that the policies have permitted an orderly adjustment to drastic change.

US regional policies and mining areas

In the huge territory of the USA, distressed areas exist in considerable variety, but regional policy has been slow to develop. Many Americans have taken the attitude that depressed areas have been found by market tests to be uncompetitive, and therefore efforts to aid them are likely to be ineffective and costly. Barry Goldwater, for example, remarked that ghost towns, like those of his native Arizona, were the necessary price of efficient national development (Cumberland, 1971). Distrust of extension of Federal power has also inhibited development of regional policy.

Many problem areas have been heavily reliant on resource exploitation, including farming, forestry and mining. President Kennedy was greatly affected by the poverty observed in West Virginian mining towns during his presidential campaign and pledged to bring to Congress a complete programme to restore the regional economy. Although steps in this direction were taken in 1961 and 1962, providing funds for public facilities in declining and stagnant communities, the most important legislation came in 1965 during the heyday of President Johnson's Great Society (Hansen, 1974). The Economic Development Administration was established to provide aid (principally through public works) to areas characterized by chronic economic distress, while the Appalachian Regional Development Act established the Appalachian Regional Commission to tackle the problems of America's largest problem region, covering an area as big as Italy, Switzerland and Austria combined and containing one-tenth of the US population.

As described earlier, in Appalachia mining and other resource-based activities had failed to sustain self-reinforcing growth. Much of the area lacked a well-developed urban hierarchy; while many towns, like Scranton and Wilkes-Barre in the Pennsylvania anthracite field, had limited scope for growth because of the mountainous terrain. Much of the region had a dense, rural non-farm population.

Isolation was a major problem, despite apparent proximity to Megalopolis. Mining damage was immense; Appalachia had 26 per cent of all US land disturbed by surface mining, and over 4,000 miles of rivers polluted by acid mine drainage.

The ARC was given funding in a number of areas, and the authority to supplement the grants made by other Federal agencies, which might not be forthcoming if matching funds could not be provided locally. It has a pump-priming role. Its policies have been criticized for their great emphasis on highway construction, which took over 75 per cent of the initial funds. The aim was to reduce isolation from the national market and allow people to commute to job opportunities and facilities. The programme has been defended as providing the vital matrix without which the effectiveness of other investments would be greatly reduced (Newman, 1972). Moreover, highway development must be seen as only the first phase. Emphasis in policy has shifted subsequently—first to 'human resources development' (health, education, etc.) and secondly to 'enterprise development' (investment in more direct support for the private enterprise—access roads, water, industrial sites, etc.). This is intended to produce a progressive removal of the various obstacles to growth—isolation, poor quality population, lack of economic overhead capital.

Although other regional commissions were created in 1965 and since, and legislation was being drawn up in 1979 to extend them across the whole nation, none have remotely matched the funding and powers of the ARC, which remains the outstanding single feature of US regional policy, and one which is already claiming significant impact on Appalachian fortunes. Contrast with British policy, with its emphasis on direct inducement to private firms, is considerable. The stress on cost effectiveness is also more explicit in Appalachian policy. The 1965 Act specified that investments should be 'concentrated in areas where there was the greatest potential for growth, and where the expected return on public dollars would be greatest' (colloquially known as the 'bang for the buck'). The emphasis is thus on potential as much as need, and one corollary is the development of an explicit growth centre approach.

5 'Late' stage development—the mining revival and the return of the resource frontier

No precise delimitation is possible between 'middle' and 'late' stage development. The nomenclature adopted is intended merely to provide a broad framework suggesting the principal associations between development stages and mining activity. In this chapter the recent experience of the UK is used to discuss the possible emergence of a new relationship in the 'late' stage. Depressed area problems and regional policy are still prominent but a number of factors have combined to stimulate a revival of interest in indigenous mineral sources.

Factors in the UK mining revival

(a) *Mineral prices*. Of fundamental importance has been a set of factors governing the global availability and pricing of minerals. Until the 1960s the long-term price trend had been stationary or sloping gently downwards, reflecting both the substitution possibilities for different minerals and the downward cost trend resulting from technological change (Bosson and Varon, 1977). This trend had been to the disadvantage of regions reliant upon mineral exports. In the 1960s, however, the cost trend for many minerals began to turn upwards as economies of increased scale of production became exhausted. Prices of some minerals began to rise erratically, the changes culminating in the chaos induced by the quadrupling of world oil prices in 1973–4. According to Bosson and Varon (1977) prices of minerals and metals did better than other product groups and improved in real terms between 1960 and 1973. In the recessionary climate since 1974 prices of many minerals have been depressed, though those of metals remained highly volatile. Tin, for example, advanced in price from £5,000 a tonne in late 1976 to over £8,000 a tonne in late 1978. Metal prices tend to be cyclical and are affected by general economic conditions, supply bottlenecks, the ability of producers to organize themselves, and, sometimes, US stock-piling policy.

The oil price rises induced by OPEC in the last decade have increased the problems of cost inflation in many mining industries, but have stimulated the search for *energy* minerals. High prices have ensured the commercial viability of high-cost oilfields in the North Sea; coal has been given the opportunity to arrest a long period of decline'; the fillip to nuclear power prospects has intensified the hunt for uranium.

(b) *The quest for safe havens*. Stimulus to focus mineral exploration in the *advanced* countries has come from the increasingly hostile condition facing multinational companies in many developing nations. Events in Iran in 1979 exemplify the problem. Many companies have turned from zones of political instability to safer havens. A growing proportion of world mineral exploitation in the last two decades has been taking place in the developed lands, especially in Australia, North America and South Africa, but also the UK. Some international corporations have also responded to the pressures by attempting to diversify and increase their stake in indigenous enterprise. For example, Consolidated Goldfields, besides developing the Wheal Jane tin-mine in Cornwall, has absorbed the Amey Roadstone Corporation.

Balance of payments problems, intensified by reliance on imported minerals, have led the UK government to react favourably towards indigenous development, particularly in the fuel sector. Oil companies seeking to develop North Sea fields have been accorded considerable tax concessions on exploration and development costs. The almost total dependence of British industry upon foreign supplies of non-ferrous metals (Zuckerman Commission, 1972) was undoubtedly a factor behind the Mineral Exploration and Investment Grants Act of 1972. This introduced an incentive scheme, whereby 35 per cent of the cost of mineral prospecting and evaluation for non-ferrous metals, barium and potash, could be met by a government grant, repayable only if the prospect developed into a productive mine. By 1979 a total of £2.6m. had been granted, and 48 companies had applied for aid for 174 projects. Although the scheme is nation-wide, most activities are taking place in Scotland, northern England, Wales and the south-west peninsula. These are the areas with the appropriate geology, as well as, by coincidence, lying largely within the assisted areas of regional policy and in upland areas of outstanding landscape quality. Nevertheless this scheme is of a modest scale, and in the view of many critics has hardly compensated for inadequacies in mineral rights legislation, and the tax system, despite modification in 1970, is still considered unfavourable to mining.

(c) *The influence of regional policy*. In the same period, regional policy changes have brought some limited assistance. Since 1960 regional policy has been broadened from a narrow dependence on the manufacturing sector. The 1966 Industrial Development Act made available investment grants on mineral exploration and development, initially at a nation-wide rate of 20 per cent, doubled in the Development Areas. Investment grants were abandoned by the Conservative government in 1970, and replaced by a system of depreciation allowances and tax concessions. The change was unwelcome to the mining industry as grants had the effect of improving cash flow during the development stages, which the new measures did not. Cleveland Potash Ltd., developing the Boulby potash mine, were forced by the withdrawal of grants to find an extra £6m.

The new system was short-lived. In 1972, with one of those abrupt reversals of policy that characterized the Heath administration, regional development grants were introduced, for which mining industries qualified. These grants covered capital expenditure on new plant and machinery and on mining works, at rates of 22 per cent in SDAs and 20 per cent in DAs. The change provided some compensation for the loss of investment grants in 1970. 'Selective assistance' was another important aspect of the 1972 legislation for which mining qualified, enabling the government to aid projects in the assisted areas which provided additional employment, and to safeguard existing employment. In 1973, for example, the government made loans to preserve jobs at the new Cornish tin-mine, Pendarves.

However, the benefits of the 1972 policy also proved brief in duration. Always one of the more marginal of regional policy provisions, the eligibility of mining for regional development grants was withdrawn in 1976-7 in a package of public expenditure cuts, though much oil-related development continued to qualify. The change was justified in a statement that mining location was not for the most part determined by any incentive, a facile observation ignoring the fact that most zones of geological interest for several important minerals lay in the assisted areas. Mining industries were still eligible for selective assistance, but in the shift to a more rigorous regional policy announced in 1979, this is to become extremely selective. In compensation, British membership of the EEC has brought advantages, through the ECSC (which had contributed £340m. to coal's investment projects by 1978) and the European Investment Bank.

Although regional aid to the mining industry has been essentially modest, and has been motivated as much by trade as by regional considerations, it has helped to provide a more favourable environment for mining exploitation. However, during the same period strong countervailing forces have also developed.

Environmental opposition

As societies become more affluent, demand for consumer goods slackens in favour of the 'superior' goods of amenity resources and environmental quality, and as the latter becomes more scarce it becomes more precious (O'Riordan, 1971). Society has become less tolerant of the 'externalities' produced by mining operations—polluted land, air and water, spoliation of scenic beauty, increased noise and traffic. Proposals to exploit minerals in many areas therefore meet increasingly organized protest, though most environmentalist groups remain peripheral to the political power base.

It might be argued that the UK has a well-developed physical planning system, which provides explicit controls over mineral working, and indeed in the case of some minerals (such as salt, ironstone, coal) there is special legislation imposing high compensation and restoration standards. However, many environmentalists doubt the adequacy of protection under the planning system with regard to major mineral projects and believe, with some justification, that the system of planning inquiries is weighted in favour of the developer. The Planning Inquiry Commission machinery established in 1968 lies unused, while environmental impact statements, despite the recommendations of Catlow and Thirlwall (1976), are not mandatory. Furthermore, doubts about the real strength of government support for environmental protection have been increased by the haste with which it has rushed to aid investment in domestic energy supply since 1973. In 1975, for example, the Secretary of State for Scotland was given power to compulsorily purchase any land urgently needed for North Sea oil without public inquiry.

The conflict between conservationists and mineral developers has gained special impetus from the location of prized minerals within National Park boundaries. Among many controversies the following are notable. In Snowdonia, RTZ (Rio Tinto Zinc) gained approval in 1971 for prospecting for gold in the Mawddach estuary and copper at Capel Hermon but abandoned the prospects without making further applications. Three applications for potash mines in north Yorkshire gained planning permission in 1968 and 1970, though only one mine was sunk because of a glut in the potash market. However, an extension of planning permission on Whitby Potash's projected solution mine was refused in 1979 after a further public inquiry. In 1971 the application by English China Clays to extend its Lee Moor open-pit workings within the Dartmoor National Park boundary won qualified approval.

The debate in some parks has been complicated by their lack of other economic opportunities; many have suffered from unemployment and depopulation. Many also have long traditions of mining and quarrying, 'there is scarcely a hill in North Wales without a hole in its side' (Searle, 1975). Local people and their elected representatives (of varied political persuasion) therefore frequently favour mineral development despite adverse environmental consequences. 'Jobs before scenery' is a familiar cry. Opposition to mining is often strongest from outsiders who use the parks for recreation, second homes and retirement, and from national organizations like the Ramblers' Association and the Friends of the Earth (Robertson, 1974).

Such conflicts are not limited to National Parks. As long ago as 1960 an application to disturb the 'homely and tranquil beauty' of north Oxfordshire for ironstone was rejected because of the implied damage to amenity (Gregory, 1971). In the 1970s, NCB plans to mine the farmlands of lowland England have aroused a furore in the Vale of Belvoir. Here the growing dormitory role of some settlements for near-by towns and cities adds the protest of commuter to those of landowner, farmer and conservationist (North and Spooner, 1978a).

The renaissance in mineral exploitation in the 1960s and 1970s has not therefore been unopposed, although opposition to exploitation of *energy* minerals was muted in the mid-1970s in the gloom generated by the oil crisis. At the 1975 inquiry into the proposed Selby coal-mines, the inspector reported that he could detect 'no outright objection to the grant of planning permission to the applications and the principle that coal contained in the site should be worked in the national interest was accepted by all witnesses' (Department of Environment, 1976). Such patriotism has since rapidly dissipated. The most spectacular developments have nevertheless been in the energy field. Developments among other minerals have ultimately proved more modest. The major events have been the opening of three new tin-mines in Cornwall, though all have struggled to survive, and the Boulby potash-mine in north Yorkshire. In the Pennines a number of investment schemes have been implemented for fluorspar.

A new stage in the mining cycle

The evidence supports the argument by Robertson (1974) that Hewett's national mining cycle must be extended to a sixth phase, in which advanced technology is utilized to exploit deep and perhaps low-grade deposits, though the argument is greatly strengthened by inclusion of the energy minerals. Politics play an increasing role, both internationally in influencing world prices and the security of overseas sources, and internally in the reaction of society to the environmental implications of increased production.

If we consider the typology of regional development, the extent of metropolitan influence in modern Britain makes identification of 'resource frontiers' according to Friedmann's original type impossible. Colonization of wilderness cannot occur. However, through the accident of resource endowment, many developments are occurring in the peripheral regions of the UK, giving them potential significance in reshaping regional patterns and redressing regional imbalances. In underdeveloped rural regions like the Scottish Highlands hopes have been raised that mineral exploitation might provide a new type of growth point leading to structural and spatial reorganization and pushing the region on to a path of cumulative growth. To explore further the role of mining expansion in modern regional development in the UK, new developments in oil and natural gas, coal and tin are now examined. These are occurring in a variety of regional contexts, including the redevelopment of mining regions (coal, tin), the shift of mining into contiguous zones along a frontier (coal), and the establishment of new enclaves of activity in areas with little previous association with mineral extraction (oil and gas).

North Sea oil and gas

Undoubtedly the outstanding new feature of the economic geography of UK mineral production since 1960 has been the development of offshore oil and gas resources and its related onshore impact. Exploration began in the early 1960s in the southern North Sea, following the discovery of natural gas in the Dutch Slochteren field. The pace of development was rapid, and five fields were in production by 1972. Oil was struck further north in the Forties field in 1970, and first landed in 1975; by 1979 eight other fields were in commercial operation as well as the Frigg gas field; total annual oil production had reached 54m. tonnes and self-sufficiency for the UK was in sight. While the early gas developments centred on Great Yarmouth, oil's impact was felt much further north, in Teesside and the Firth of Forth, but more especially in north-east Scotland around Aberdeen and Peterhead, in the Moray and Cromarty Firth areas, and in Orkney and Shetland.

The onshore impact has taken several forms. Service centres for offshore operations were created, both 'advance' bases like Peterhead and Lerwick, and 'major' bases like Great Yarmouth and, above all, Aberdeen (Chapman, 1976). Terminal facilities have been built, as at Cruden Bay, Bacton, St. Fergus and Sullom Voe.

Concrete and steel production-platform sites have proliferated, some in remote locations providing the necessary deep-water access. Other manufacturing industries have developed, including steel pipe, turbine and pump-making. On the Cromarty Firth a new oil refinery is being built.

Development of each field follows the discovery-depletion cycle. Government energy policy is based on the assumption that by 1990 the production profiles from the first 14 fields will have passed their peak, and that total supplies will decline before 2000 (Secretary of State for Energy, 1978). In the early 1980s, production should reach between 100 and 150m. tonnes p.a. Development is capital-intensive and each phase of the cycle makes variable demands on employment. Recent predictions suggest that total employment creation will not peak before 1980.

A 1974 survey revealed that, excluding construction, a total of 26,000 oil-related jobs had been created in Scotland, 50 per cent of them in new units, and 75 per cent in manufacturing (Pounce, Upson and Walker, 1976). Not all these were new jobs, however, in the older industrial areas many represented redeployment. By 1976 estimates indicated that, including construction and offshore installation, the total oil-related employment had risen to 44,000. Assuming a multiplier in the range of 1.2 to 1.4 the estimated total employment might lie between 55,000 and 64,000 (*Scottish Economic Bulletin*, 1977). Much depends on the multiplier and the ratio of total effects to direct effects alone. Mackay and Mackay (1975) argued that the size of the multiplier was unlikely to exceed 2, but did concede that there was a possibility that this did less than justice to the power of North Sea oil to create a new growth dynamic—a 'super-multiplier' to sustain expansion. The potential excitement and psychological boost created by the industry makes prediction hazardous. Whatever the precise value of the multiplier, it is also clear that it will display considerable spatial variation within Scotland (Chapman, 1976).

Viewed at the macro-scale, the number of jobs created is puny in comparison with the needs of Scotland as a whole. Between 1966 and 1973 the primary and secondary sectors were losing 25,000 jobs p.a.; Mackay and Mackay (1975) concluded that there were distinct limits to the regeneration that could be achieved, and little prospect that the direct benefits from oil would be large enough to produce a major and lasting change in Scotland's economic performance. Keeble (1976) doubted the ability of the oil and gas-related developments to have more than a minimal and local impact on the geographical distribution of *manufacturing* within the UK *as a whole*. Experience with gas in eastern England might be indicative;

there the major industrialization envisaged by Odell (1966) had not occurred. The transportability of gas by pipeline, pricing policies that did not discriminate in favour of the landing zone, and the inertia of installed facilities elsewhere, had combined to limit industrial development. Within Scotland, oil's impact on industrial location has remained relatively small; most of the output flows to existing industrial complexes like that at Grangemouth.

The regional impact of oil-related development has been viewed as unsatisfactory *within* Scotland. West-central Scotland contains the biggest single concentration of urban deprivation in the UK, but this area remains remote from much of the oil activity. Between 1971 and 1980, 153,000 to 236,000 job losses were expected in this area, dwarfing the 13,900 direct oil-related jobs gained by 1974. Lewis and McNicol (1978) stress that the chief importance of oil-based development in this region may be in promoting technological change in existing industry. Certainly the oil industry appears impotent to rejuvenate the old industrial centres by direct job creation.

However, with respect to northern Scotland, itself a problem region of a different type, there are some grounds for optimism, despite considerable social and environmental stress. Lewis and McNicol (1978) see the oil-based developments not as a panacea, but as a means of stimulating at least part of the industrialization process that is needed. Already at least 10 per cent of Highland employment is oil-related, and locally, in the Shetlands and in Cromarty, economic conditions have been transformed. They argue, perhaps optimistically, that the infrastructural investment (as in transport systems) already induced by oil-based development may in turn stimulate other types of industrialization.

Moreover, the clustering of oil-related development in favoured locations is aiding the implementation of growth-centre policies, for long dear to the heart of regional planners. As long ago as 1948 the coastal strip around the Moray and Cromarty Firths was selected as a DA because of its suitability as a focus for industrial development in the Highlands, and this area has featured prominently in HIDB (Highlands and Islands Development Board) strategies.

Elsewhere on the east coast, at Aberdeen, the strong development of services has produced another focus for growth, resulting in its upgrading from DA to IA status in 1977, a response to its economic buoyancy and dwindling unemployment. By 1982 it will lose assisted-area status entirely, as will Peterhead, Shetland and Orkney.

Strong *local* growth foci have emerged in this modern resource frontier region, but it is difficult to disagree with Mackay and

Mackay's conclusion that the major benefits arising from North Sea oil lie in the revenues accruing to the national government rather than in direct employment and income creation. Such revenues, from royalties, petroleum tax and corporation tax, are expected to rise to £3.5 billion per year in the mid-1980s. The use of these revenues in a regional context is crucial; at present the money enters the general accounting system. The real impact from North Sea oil in regional development terms would come if such money was channelled directly into the rejuvenation of Clydeside or into the oil-producing region. Disposition of the oil money remains a highly political decision. In the long run, much may depend upon how far the oil-producing regions can develop activities which can continue to serve markets elsewhere as the oilfields pass their peak.

Coal

In contrast with the oil developments, changes in the coal industry taking place in the late 1970s follow long-established trends, both at the inter- and intra-regional scales. Under the 1974 Plan for Coal a threefold programme is being carried out to stabilize total coal output at 135m. tons p.a. (120m. deep-mined), entailing investment in 42m. tons p.a. new capacity by 1985 to replace that being lost at exhausting collieries. 20m. tons were to come from new collieries, 13m. tons from major reconstruction at long-life collieries, and 9m. from life-extension at short-life collieries. Subsequently these figures have been revised because of delays in obtaining planning permission, and now only 10m. of the 120m. ton deep-mined target for 1985 will be obtained from new mines.

The NCB has made out its case for a much larger expansion to 170m. tons (150m. underground) by the year 2000 to meet projected demand growth; this might imply the sinking of 30 new mines (Department of Energy, 1977). Sufficient reserves have been discovered to support this scale of expansion. Some critics, however, question the ability of the industry to market even the 1985 target, pointing to stagnation in the vital electricity and steel markets, as well as growing competition from low-cost competitors overseas (Manners, 1976).

Government policy clearly appreciates the need to make the industry price-competitive partly by removing the burden of uneconomic collieries in the older fields, but implementation of colliery closures is often difficult to achieve in the face of opposition from a strongly organized workforce. The NCB case for expansion is based largely upon *long-term* expectations of market growth, emphasizing expansion of direct industrial and domestic use, as new technology gives commercial viability to such processes as fluidized bed combustion, liquefaction, and manufacture of substitute natural gas (SNG).

The trend towards concentration of production in the more 'central' coalfields of the Yorkshire and East and South Midland regions (Frontispiece), is being intensified by the regional distribution of investment under the Plan for Coal. The expensive Selby project is a major ingredient, but reconstruction schemes at existing collieries also favour these regions, where both mining and market conditions are generally most propitious. In 1978 a major scheme for reorganization of the Barnsley district of Yorkshire was announced, to bring almost all of its 8m. tons output to the surface at only three major outlets. This district also has small new drift mines at Royston and Kinsley. Of the 21 most expensive schemes at existing collieries announced by the end of 1976 only five were in 'peripheral' fields.

Areas like South Wales and the North-East have long enjoyed advantages in the production of special coals (anthracite in South Wales, coking in both) for regional market outlets, and both areas can boast a number of impressive investment schemes, including the small anthracite-mine at Betws. However, the relatively high cost of production in South Wales especially and the developing weakness of the steel industry do not augur well for the long-term future of mining in these areas. In Scotland the situation is also grave; the fields have no special coals and alternative sources of energy are making large inroads into the electricity market. The regional problems of these peripheral coalfields are thus unlikely to be alleviated by the changing fortunes of the coal industry.

The long-standing trend of intra-regional shift towards deeper concealed seams has also received fresh momentum. In the great Yorkshire-Derbyshire-Nottinghamshire field the eastward movement of the mining frontier, static since the early 1960s, is being resumed. Already the major salient of the Selby coalfield is being developed, a massive project to produce 10m. tons of coal p.a., utilizing a drift and five satellite mines.

Immediately to the south, coal has been found around Snaith, and Thorne Colliery, which closed in 1956 because of water problems, is being reopened. Other finds have been made along the eastern fringes of the field as in the Vale of Witham, but the southern counterpart of Selby is the north-east Leicestershire project, part of which includes the Vale of Belvoir. Here a decision on plans to extract 7m. tons p.a. from 3 mines is awaited. This frontier zone of the coalfield has the major advantage of favourable geological conditions for large-scale high productivity mining, but also of short hauls to established markets in the Aire and Trent power stations. The same pattern of outward movement to deeper seams is repeated on a smaller scale in the Cannock coalfield, where plans have been put forward for the 2m. tons p.a. Park colliery, east of Stafford.

In the regional context, the investments in the central coalfields bring greater stability of employment, possibilities of labour recruitment and substantial regional expenditure. However, the intra-regional shifts bring sub-regional problems. With the exception of Thorne, most parts of the frontier zone have been relatively free of unemployment problems, and in the case of Selby, it has enjoyed a healthy diversity of small-scale economic activity (North and Spooner, 1976). The provision of 4,000 jobs at Selby and 3,800 at Belvoir, plus multiplier effects, brings therefore a risk of strain on local labour markets. Local employers' attitudes have been ambivalent, some fearing competition for scarce labour, others welcoming the augmentation of the female labour pool that might arise from immigration of miners' families.

Furthermore, fears have been expressed that the extension of the mining frontier might lead to an acceleration of withdrawal from the older mining districts to the west, adding to these areas considerable unemployment problems, and wasting community investment. A more realistic viewpoint has been adopted in the Belvoir debate by the mineworkers of the north-west Leicestershire coalfield, where all mines will close anyway within 20 years. They have formed a vigorous lobby in favour of the Belvoir project. In the plans for the Park colliery a further stage in this argument has been reached, with the NCB citing as a positive advantage of the project the opportunity to retain the skills of mineworkers employed at Staffordshire collieries, even specifying an individual colliery, West Cannock No. 5, where the impending exhaustion of reserves will supply nearly half the workforce needed.

Partly because of the lack of *local* pressure for new jobs in Selby and Belvoir, much of the debate around these proposals has centred upon issues relating to the physical and social environment (North and Spooner, 1977).

Compared to the oil industry, the direct impact of coal on regional employment is large, though smaller than in the past. In the three 'central' regions for example, nearly 130,000 wage-earners were employed in 1978; the total of coal-related jobs are likely to approach 200,000, even using a fairly modest multiplier. In the USA Miernyk (1975) utilized a 1.9 multiplier for coal. Nevertheless the investment schemes are unlikely to do more than stabilize the industry's labour demands in these regions. Compared to oil, transport problems suggest a greater utilization of the coal within the production zone, but while the major use of coal remains electricity generation, the impact on industrial location will continue to be small.

Tin

Although tin remains the UK's major non-ferrous metal mining industry, only two mines existed in Cornwall in 1960, where less than one hundred years previously there had been 300. The decline of the industry was not due to the exhaustion of the ore deposits, but rather to internal weaknesses in the industry and to the rise of low-cost overseas producers. In the 1960s, however, changing market conditions, political uncertainties in South-East Asia, and modest government encouragement led to a wave of exploration in south-west England, which culminated in the opening of three new mines in west Cornwall between 1971 and 1976: Pendarves, Wheal Jane and Mount Wellington. The impact of the new mines (especially Wheal Jane), plus new investments at Geevor and South Crofty, was to push the output of tin-in-concentrate from 1,722 tonnes in 1970 to nearly 4,000 tonnes in 1977, when employment in the industry exceeded 1,500 workers. The expansion has been aided by regional policy; Mount Wellington, for example, a £5.3m. project, received £0.8m. in regional development grants and £0.8m. in government loans. Withdrawal of grants in 1977 has caused difficulties.

Each new mine has had a chequered history. Pendarves went bankrupt in 1973 but was taken over by St. Piran, who treat its small output at their South Crofty mine. Steep cost inflation and price recession led to operating losses at Wheal Jane in 1975 but fortunately prices rose rapidly thereafter. However, in 1978 a bombshell struck the re-emergent industry. Mount Wellington, still not yet in full production, closed, allegedly because of poor grades and water problems; this precipitated closure at the adjacent Wheal Jane, where the additional water problems arising from Mount Wellington's closure was the final blow, on top of a history of disappointing grades. The real cause, however, of Wheal Jane's failure may have been mismanagement of the project by Consolidated Goldfields (Blunden, 1979). A further twist to the saga came in 1979 with the announcement that Wheal Jane and Mount Wellington are to be taken over and reopened by an international mining group, including RTZ.

West Cornwall remains a problem area, with its economic structure heavily weighted towards agriculture and tourism. Nine-

52 Mining and regional development

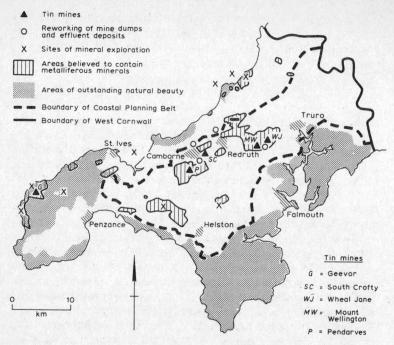

Fig. 2. Tin in west Cornwall in the 1970s. N.B. Some of the mines, as well as the small effluent workings, have not operated continuously. *Sources*: various, including Cornwall County Council (1970): West Cornwall Study.

teenth-century mining has left a chaotic settlement pattern and scarred the landscape and the economy. Unemployment rates are high. Manufacturing industry is limited and was dominated until recently by two large enterprises, one of which, the Falmouth ship-repair yard, has now severely contracted. An inflow of new firms under the stimulus of regional policy has provided some respite, but a high proportion of jobs created are female. The tourist industry has proved disappointing as a source of new employment. National and local government policies have aimed to stimulate employment growth, particularly in a 'restraint free' corridor stretching from near Penzance to Truro, away from the coastal planning belt (Cornwall C.C., 1970), (Fig. 2).

In such a context, the possible contribution of mining to regional regeneration is important, not least in the psychological boost engendered by pride in reliance on indigenous resources. Data on the impact of the mining sector are limited. In 1975 the industry expended £4.8m. on wages and salaries, and other local payments totalled £2.9m. (Cornish MDA, 1976). Blunden (1979) argues from Canadian experience that the total job creation from the 1,500 direct jobs *may* be in the order of 7,500, a remarkably strong multiplier effect if true. Wheal Jane, with 450 male and 10 female workers in 1976, was the largest surviving *male* employer to have set up in Cornwall since 1945. In an area where male employment is scarce, the mine's value needs no underlining. Furthermore, although a nucleus of skilled workers was imported and housed by the company in Truro, Consolidated Goldfields claimed a high proportion of its workforce was recruited locally. Surprisingly, however, periodic difficulties have been faced both here and at Geevor with labour recruitment. On the negative side, the industry still retains many of the classic features of an exploitative industry operating in a peripheral region; most of the companies are based outside the region and the entire output of tin is shipped for processing elsewhere.

In the past some problems have been encountered with clash with amenity (Goodridge, 1966), but this seems unlikely to be a major restraint on future expansion. The major difficulties lie with the industry itself, the cost of operation and world price levels, and these problems are clearly considerable. Tin's hold in south-west England remains precarious.

A new role in regional development?

This chapter has examined the UK mining revival, the problems that it is facing and the new patterns that it is creating. It is worth stressing that the problems are by no means unique to the advanced economy of the UK. In the USA, the large-scale expansion of *surface* coal-mining in western states like Wyoming is opposed by environmental groups and local agricultural interests, threatened by land despoliation and 'potentially undesirable social and economic effects in a primarily rural, sparsely populated, agriculturally oriented society' (Griffith and Clarke, 1979). A particularly difficult problem in some regions is the conflict between mineral developers and ethnic minorities with a different land ethic. In Queensland, for example, further Aboriginal tragedies are in the making, in the quest for bauxite and uranium.

Given the evidence presented, what conclusions can be drawn about the role of modern resource frontiers in advanced economies like the UK? The evidence suggests the need for caution. In the case of the metals and localized non-metals, the direct local impact of individual projects is often considerable and should not be underestimated, but the nature of developments that have occurred appears to conform largely to a 'hinterland' style. The continuing volatility of market conditions makes much development marginal. 'The cynic may see the current activity as yet another flutter of a new generation of moths into the old candle-flame', thus was the Cornish tin revival described by the historian, Barton, in 1967, and events in the 1970s have come close to justifying his words.

The case of the energy minerals merits closer analysis, partly because of the far greater scale of development involved, partly because of the traditional role of such industries in industrial location, partly because the era of cheap energy has ended. Is a new relationship between energy minerals and regional development in prospect?

The key to this question may lie in the newly emerged tendency for energy prices to rise more rapidly than the general price level. One potential consequence of this is a shift of real income from energy-consuming to energy-producing regions. Rising energy prices might alter the distribution of regional purchasing power. Miernyk (1977), from US experience, argues that we have become too used to thinking of energy-producing states as characterized by low *per capita* incomes. He points out that one effect of rising energy prices during the 1975 recession was a differential impact on the fiscal position of states. Problems were exacerbated in energy-consuming states, while the recession was cushioned in energy-producing states, some like Wyoming enjoying near-boom conditions. In some states large budget surpluses were mainly attributable to unanticipated growth of coal tax revenue. Similary he argued that the coal boom was the most important cause of rapid personal income increases and falling unemployment in parts of Appalachia (Miernyk, 1975).

Miernyk's argument is a fascinating one; it suggests a possible long-run shift in the balance between an energy-poor heartland and an energy-rich hinterland. The extent to which this will occur depends upon a complex of factors determining the geographical destination of the 'windfall profits' created by the high energy prices. Much of this windfall may be dispersed by the pattern of ownership in the energy industries—or even channelled back to the heartland. However, much should remain in the producing area— in higher wages and rents, and through the incentive to companies to plough back investment into further energy production. A crucial factor may be the extent of the local or state taxation take; some states have 'severance' taxes on coal production (in Kentucky and West Virginia this amounts to $c.4$ per cent of the value of coal produced). Another crucial factor will be the extent to which the energy sources, as they become dearer, can again exercise an attractive force over the location of consumer industries. This will depend upon the extent of spatial variation in energy costs. In this respect Miernyk's argument appears weakest. The tranportability of oil, gas and electricity, and even of coal by slurry pipelines, zonal or blanket pricing policies on electricity, the entrenched advantages of metropolitan industrial regions, will all tend to work against a shift to the energy-producing regions. Much depends on the impact of new technology. Miernyk (1975) points out that the development of low-Btu gasification could be crucial, as this gas will have to be burned close to its source of production, and also reminds us that the cost of *transporting* energy will rise.

Although few would be bold enough to deny that the era of cheap energy is over, the evidence for shifts of the type postulated remains limited and the arguments hypothetical. Analogies between American and British experience are dangerous because of the very different size of area and political framework. The experience to date with North Sea oil and gas suggests that analogies to French experience with the Lacq gasfield may be more pertinent. When this field was developed in the 1950s hopes were high that it would become a 'growth point' for the backward Aquitaine region. By the 1970s the field was supplying 80 per cent of French gas needs and had become the site of an industrial complex utilizing the gas both as a fuel (electricity generation, and thence aluminium smelting) and as a raw material for petro-chemicals. However, Penouil (1969) considered that the 'growth point' had proved to be an illusion. Certainly Lacq had played the role of a *local* growth focus and had had an impact on part of Pyrénées-Atlantiques, but it had not acted as a *regional* growth point. Only 25 per cent of the gas is consumed in south-west France. A close relationship had not formed between the industrial complex and the regional environment. Lacq had remained a cathedral in the French desert.

Hansen (1968) went further in citing Aydalot's analogy between Lacq and the colonial enclaves of the Third World, pointing out that the companies involved were Paris-based, that the work force was small and imported, while Aquitaine had the extra disadvantage of not being able to tax the large firms.

This analysis may overstate the Lacq argument and refers to the era of *cheap* energy, but it suggests the need for caution in assessing the role of new energy sources in UK regional development. Certainly *local* growth foci have been created but the existence of a major regional growth point is still questionable. The argument of Mackay and Mackay (1975) that the major benefit of North Sea oil is *national* is a powerful one, and suggests no fundamental alteration in the national centre-periphery structure without deliberate policy measures. Much of the windfall income arising from high energy prices accrues to multinational companies and the national government. The use of government revenues in a regional context is crucial. The possible 'trend-breaking shift' in Western European industrial location hypothesized by Odell (1978) may await political and institutional change.

From the viewpoint of peripheral energy-producing regions, the question must be whether they gain sufficiently from the rise in energy prices, or whether the bulk of the effect is leaked and continues to benefit the centre. A secondary question is whether such regions with their often fragile communities and environment, are adequately compensated for social and environmental disruption.

A serious issue must therefore be the direct take by local government of a share in the energy revenue, over and above that achieved by the rating system. In the UK one example already stands out of what might be accomplished in this direction. In 1974 the Shetland local government achieved considerable control over oil-related development and social and environmental disruption with the passage of the private Zetland County Council Act. This gave compulsory purchase powers for any land selected for oil-related development and established a reserve or 'disturbance' fund from a special tax, linked to world price and company profitability, on all oil entering the islands' facilities. The fund should amount to £6–30m. over the next 10–15 years, and is effectively a local land development tax paid by the oil beneficiaries to the local population (O'Riordan, 1977). This fund will internalize to the Shetlands some of the windfall income generated by high oil prices. In the USA, the state of Kentucky instituted in 1974 a 'coal county fund', whereby half the surplus revenues accruing from coal severance tax (in excess of state budget estimates) were reserved for use in the counties that actually produced the coal, in proportion to their contribution to the tax, rather than being dispersed over the whole state. If the new resource frontier regions of the late twentieth-century UK are to gain adequately from the fruits of energy production in an era of expensive energy, the examples of Shetland and Kentucky may be highly relevant.

References

Abler, R., Adams, J. S., and Gould, P. (1971) *Spatial Organisation*, Englewood Cliffs, N. J.
Adler, J. H. (1961) 'Changes in the Role of Resources at Different Stages of Economic Development', in Spengler, J. J. (ed.) *Natural Resources and Economic Growth*, Washington, D.C.
Baldwin, R. D. (1966) *Economic Development and Export Growth: a Study of Northern Rhodesia 1920–60*, Berkeley and Los Angeles.
Barton, D. B. (1967) *A History of Tin Mining and Smelting in Cornwall*, Truro.
Berry, B. J. L. (1973) *Growth Centres in the American Urban System, Volume One*, Cambridge, Mass.
Blakemore, H. (1971) 'Chile', in Blakemore, H. and Smith, C. T. (eds.) *Latin America: Geographical Perspectives*, London.
Blowers, A. (1972) 'Unit 12, Social Planning, Open University course D281', *New Trends in Geography*, Milton Keynes.
Blunden, J. R. (1975) *The Mineral Resources of Britain*, London.
—— (1977) 'Units 4–6 Man's Impact on the Environment; Resource Exploitation, Open University course D204', *Fundamentals of Human Geography*, Milton Keynes.
—— (1979) *Cornish Tin: Future Prospects for the Exploitation of an Internationally Traded Commodity in terms of its Impact on the Regional Economy*, Regional Studies Association, London.
Bosson, R. and Varon, B. (1977) *The Mining Industry and the Developing Countries*, Oxford.
Boudeville, J. R. (1966) *Problems of Regional Economic Planning*, Edinburgh.
Brookfield, H. (1975) *Interdependent Development*, London.
Brown, A. J. (1972) *The Framework of Regional Economics in the United Kingdom*, Cambridge.
—— and Burrows, E. M. (1977) *Regional Economic Problems*, London.
Brown, M. and Webb, J. N. (1941) *Seven Stranded Coal Towns*, Washington, D.C.
Bulmer, M. (1978) *Mining and Social Change*, London.
Caesar, A. A. L. (1964) 'Planning and the Geography of Great Britain', *Advancement of Science*, **21**, 91, 230–40.
Catlow, J. and Thirlwall, C. G. (1976) *Environmental Impact Analysis*, Department of the Environment, Research Report 11, H.M.S.O., London.
Caudill, H. M. (1962) *Night comes to the Cumberlands, the Biography of a Depressed Area*, Boston.
Chapman, K. (1976) *North Sea Oil and Gas*, Newton Abbot.
Chinitz, B. (1960) 'Contrasts in Agglomeration: New York and Pittsburgh', *American Economic Review*, **51**, 2, 279–89.
Cornish Mining Development Association. *Annual Reports 1967–*
Cornwall County Council (1970) *West Cornwall Study*, Truro.
Cumberland, J. H. (1971) *Regional Development Experiences and Prospects in the U.S.A.*, The Hague.
Department of Economic Affairs (1969) *The Intermediate Areas: Report of a committee under the Chairmanship of Sir Joseph Hunt*. Cmnd 3998, H.M.S.O., London.
Department of Energy (1977) *Coal for the Future: Progress with Plan for Coal and Prospects for the Year 2000*, H.M.S.O., London.

Department of the Environment (1976) *Development of Selby Coalfield; report of Inspector* (Unpublished).
Dixon, C. J. (1979) *Atlas of Economic Mineral Deposits*, London.
Estall, R. C. (1972) *A modern geography of the United States*, Harmondsworth.
Forrest, W. (1976) 'Selby', *Colliery Guardian*.
Friedmann, J. R. P. (1966) *Regional Development Policy: a Case Study of Venezuela*, Cambridge, Mass.
—— (1973) *Urbanisation, Planning and National Development*, Beverley Hills.
Gilbert, A. (1974) *Latin American Development*, Harmondsworth.
Goodridge, J. C. (1966) 'The Tin Mining Industry—a Growth Point for Cornwall', *Trans. Inst. Brit. Geog.*, **38**, 95–105.
—— (1967) Historical Geography of the Copper Mining Industry in Devon and Cornwall, 1800–1900. Unpublished Ph.D. thesis, University of London.
Green, D. H. (1974) Information, Perception and Decision-making in the Industrial Relocation Decision. Unpublished Ph.D. thesis, University of Reading.
Gregory, R. (1971) *The Price of Amenity*, London.
Griffith, E. D. and Clarke, A. W. (1979) 'World Coal Production', *Scientific American*, **240**, 1, 28–37.
Grunwald, J. (1964) 'Resource Aspects of Latin-American Economic Development', in Clawson, M. (ed.) *Natural Resources and International Development*, Baltimore.
Hall, P. (1973) 'England circa 1900', in Darby, H. C. (ed.) *A New Historical Geography of England*, Cambridge.
Hansen, N. M. (1968) *French Regional Planning*, Edinburgh.
—— (1974) (ed.) *Public Policy and Regional Economic Development*, Cambridge, Mass.
Hay, A. M. (1976) 'A Simple Location Theory for Mining Activity', *Geography*, **61**, 2.
Hewett, D. F. (1929) *Cycles in Mineral Production*, Technical publication No. 183 of the American Institute of Mining and Metallurgical Engineers, New York.
House, J. W. and Knight, E. M. (1967) 'Pit Closure and the Community. Report to the Ministry of Labour', *Paper on Migration and Mobility in Northern England*, No. 5, Department of Geography, University of Newcastle-upon-Tyne.
—— Ruddy, S. A., Thubron, I. M., and Storer, C. E. (1968) 'Mobility of the Northern Business Manager', *Papers on Migration and Mobility in Northern England*, No. 8.
Institute of Geological Sciences (1978) *U.K. Mineral Statistics 1977*, H.M.S.O., London.
Jones, C. F. and Darkenwald, G. G. (1965) *Economic Geography*, New York.
Keeble, D. E. (1967) 'Models of Economic Development', in Chorley, R. J. and Haggett, P. (eds.) *Models in Geography*, London.
—— (1976) *Industrial Location and Planning in the United Kingdom*, London.
Lee, C. H. (1971) *Regional Economic Growth in the U.K. since the 1880s*, Maidenhead.
Lewis, T. M. and McNicol, I. H. (1978) *North Sea Oil and Scotland's Economic Prospects*, London.
Lindqvist, S. (1972) *The Shadow: Latin America Faces the Seventies*, Harmondsworth (English translation).
Lovins, A. B. (1977) *Soft Energy Paths*, Harmondsworth.
McDivitt, J. F. and Jeffery, W. G. (1976) 'Minerals and the Developing Economies', in Vogely, W. A. (ed.) *Economics of the Mineral Industries*, New York.
—— and Manners, G. (1974) *Minerals and Men*, Baltimore.
Mackay, D. I. and Mackay, G. A. (1975) *The Political Economy of North Sea Oil*, London.

References

Abler, R., Adams, J. S., and Gould, P. (1971) *Spatial Organisation*, Englewood Cliffs, N. J.
Adler, J. H. (1961) 'Changes in the Role of Resources at Different Stages of Economic Development', in Spengler, J. J. (ed.) *Natural Resources and Economic Growth*, Washington, D.C.
Baldwin, R. D. (1966) *Economic Development and Export Growth: a Study of Northern Rhodesia 1920–60*, Berkeley and Los Angeles.
Barton, D. B. (1967) *A History of Tin Mining and Smelting in Cornwall*, Truro.
Berry, B. J. L. (1973) *Growth Centres in the American Urban System, Volume One*, Cambridge, Mass.
Blakemore, H. (1971) 'Chile', in Blakemore, H. and Smith, C. T. (eds.) *Latin America: Geographical Perspectives*, London.
Blowers, A. (1972) 'Unit 12, Social Planning, Open University course D281', *New Trends in Geography*, Milton Keynes.
Blunden, J. R. (1975) *The Mineral Resources of Britain*, London.
—— (1977) 'Units 4–6 Man's Impact on the Environment; Resource Exploitation, Open University course D204', *Fundamentals of Human Geography*, Milton Keynes.
—— (1979) *Cornish Tin: Future Prospects for the Exploitation of an Internationally Traded Commodity in terms of its Impact on the Regional Economy*, Regional Studies Association, London.
Bosson, R. and Varon, B. (1977) *The Mining Industry and the Developing Countries*, Oxford.
Boudeville, J. R. (1966) *Problems of Regional Economic Planning*, Edinburgh.
Brookfield, H. (1975) *Interdependent Development*, London.
Brown, A. J. (1972) *The Framework of Regional Economics in the United Kingdom*, Cambridge.
—— and Burrows, E. M. (1977) *Regional Economic Problems*, London.
Brown, M. and Webb, J. N. (1941) *Seven Stranded Coal Towns*, Washington, D.C.
Bulmer, M. (1978) *Mining and Social Change*, London.
Caesar, A. A. L. (1964) 'Planning and the Geography of Great Britain', *Advancement of Science*, **21**, 91, 230–40.
Catlow, J. and Thirlwall, C. G. (1976) *Environmental Impact Analysis*, Department of the Environment, Research Report 11, H.M.S.O., London.
Caudill, H. M. (1962) *Night comes to the Cumberlands, the Biography of a Depressed Area*, Boston.
Chapman, K. (1976) *North Sea Oil and Gas*, Newton Abbot.
Chinitz, B. (1960) 'Contrasts in Agglomeration: New York and Pittsburgh', *American Economic Review*, **51**, 2, 279–89.
Cornish Mining Development Association. *Annual Reports 1967–*
Cornwall County Council (1970) *West Cornwall Study*, Truro.
Cumberland, J. H. (1971) *Regional Development Experiences and Prospects in the U.S.A.*, The Hague.
Department of Economic Affairs (1969) *The Intermediate Areas: Report of a committee under the Chairmanship of Sir Joseph Hunt*. Cmnd 3998, H.M.S.O., London.
Department of Energy (1977) *Coal for the Future: Progress with Plan for Coal and Prospects for the Year 2000*, H.M.S.O., London.

Department of the Environment (1976) Development of Selby Coalfield; report of Inspector (Unpublished).

Dixon, C. J. (1979) *Atlas of Economic Mineral Deposits*, London.

Estall, R. C. (1972) *A modern geography of the United States*, Harmondsworth.

Forrest, W. (1976) 'Selby', *Colliery Guardian*.

Friedmann, J. R. P. (1966) *Regional Development Policy: a Case Study of Venezuela*, Cambridge, Mass.

—— (1973) *Urbanisation, Planning and National Development*, Beverley Hills.

Gilbert, A. (1974) *Latin American Development*, Harmondsworth.

Goodridge, J. C. (1966) 'The Tin Mining Industry—a Growth Point for Cornwall', *Trans. Inst. Brit. Geog.*, **38**, 95–105.

—— (1967) Historical Geography of the Copper Mining Industry in Devon and Cornwall, 1800–1900. Unpublished Ph.D. thesis, University of London.

Green, D. H. (1974) Information, Perception and Decision-making in the Industrial Relocation Decision. Unpublished Ph.D. thesis, University of Reading.

Gregory, R. (1971) *The Price of Amenity*, London.

Griffith, E. D. and Clarke, A. W. (1979) 'World Coal Production', *Scientific American*, **240**, 1, 28–37.

Grunwald, J. (1964) 'Resource Aspects of Latin-American Economic Development', in Clawson, M. (ed.) *Natural Resources and International Development*, Baltimore.

Hall, P. (1973) 'England circa 1900', in Darby, H. C. (ed.) *A New Historical Geography of England*, Cambridge.

Hansen, N. M. (1968) *French Regional Planning*, Edinburgh.

—— (1974) (ed.) *Public Policy and Regional Economic Development*, Cambridge, Mass.

Hay, A. M. (1976) 'A Simple Location Theory for Mining Activity', *Geography*, **61**, 2.

Hewett, D. F. (1929) *Cycles in Mineral Production*, Technical publication No. 183 of the American Institute of Mining and Metallurgical Engineers, New York.

House, J. W. and Knight, E. M. (1967) 'Pit Closure and the Community. Report to the Ministry of Labour', *Paper on Migration and Mobility in Northern England*, No. 5, Department of Geography, University of Newcastle-upon-Tyne.

—— Ruddy, S. A., Thubron, I. M., and Storer, C. E. (1968) 'Mobility of the Northern Business Manager', *Papers on Migration and Mobility in Northern England*, No. 8.

Institute of Geological Sciences (1978) *U.K. Mineral Statistics 1977*, H.M.S.O., London.

Jones, C. F. and Darkenwald, G. G. (1965) *Economic Geography*, New York.

Keeble, D. E. (1967) 'Models of Economic Development', in Chorley, R. J. and Haggett, P. (eds.) *Models in Geography*, London.

—— (1976) *Industrial Location and Planning in the United Kingdom*, London.

Lee, C. H. (1971) *Regional Economic Growth in the U.K. since the 1880s*, Maidenhead.

Lewis, T. M. and McNicol, I. H. (1978) *North Sea Oil and Scotland's Economic Prospects*, London.

Lindqvist, S. (1972) *The Shadow: Latin America Faces the Seventies*, Harmondsworth (English translation).

Lovins, A. B. (1977) *Soft Energy Paths*, Harmondsworth.

McDivitt, J. F. and Jeffery, W. G. (1976) 'Minerals and the Developing Economies', in Vogely, W. A. (ed.) *Economics of the Mineral Industries*, New York.

—— and Manners, G. (1974) *Minerals and Men*, Baltimore.

Mackay, D. I. and Mackay, G. A. (1975) *The Political Economy of North Sea Oil*, London.

McCarty, H. H. and Lindberg, G. J. (1966) *Preface to Economic Geography*, Englewood Cliffs, N. J.
Manners, G. (1964a) *The Geography of Energy*, London.
—— (1964b) (ed.) *South Wales in the Sixties*, Oxford.
—— (1969) 'New Resource Evaluations', in Cooke, R. U. and Johnson, J. (eds.) *Trends in Geography—an Introductory Survey*, Oxford.
—— (1976) 'The Changing Energy Situation in Britain', *Geography* **61**, 4, 221–31.
Miernyk, W. H. (1975) 'Coal and the Appalachian Economy', *Appalachia*, **9**, 2, 29–35.
—— (1977) 'Rising Energy Prices and Regional Economic Development', *Growth and Change*, **8**, 3, 2–7.
Mining Annual Review (1968–79)
Morse, C. (1964) 'Potentials and Hazards of Direct International Investment in Raw Materials', in Clawson, M. (ed.) *Natural Resources and International Development*, Baltimore.
Moseley, M. J. (1973) 'Growth Centres—a Shibboleth?' *Area*, **5**, 143–50.
Moyes, A. (1974) 'Post-war Changes in Coalmining in the West Midlands', *Geography*, **59**, 111–20.
Myrdal, G. M. (1957) *Economic Theory and Under-developed Regions*, London.
Newman, M. (1972) *The Political Economy of Appalachia*, Lexington.
North, D. C. (1955) 'Location Theory and Regional Economic Growth', *Journal of Political Economy*, **63**, 3, 243–58.
North, J. and Spooner, D. J. (1976) 'Yorkshire Coal Crop from Selby Farmland', *Geographical Magazine*, **48**, 9, 554–8.
—— —— (1977) 'The Great U.K. Coalrush: a Progress Report to the end of 1976, *Area*, **9**, 1, 15–27.
—— —— (1978a) 'On the Coalmining Frontier', *Town and Country Planning*, **46**, 3, 155–63.
—— —— (1978b) 'The Geography of the Coal Industry in the U.K. in the 1970s: changing directions?' *Geojournal*, **2**, 3, 255–72.
Odell, P. R. (1963) *An Economic Geography of Oil*, London.
—— (1966) 'What will Gas do to the East Coast?' *New Society*, **188**, 5, 8–9.
—— (1973) 'Major Themes in Latin America's Economic Geography', in Odell, P. R. and Preston, D. A. (eds.) *Economies and Societies in Latin America: a Geographical Interpretation*, London.
—— (1978) 'North Sea Oil and Gas Resources: their Implications for the Location of Industry in Western Europe', in Hamilton, F. E. I. (ed.) *Industrial Change*, London.
—— and Rosing, K. E. (1976) *Optimal Development of the North Sea's Oilfields*, London.
O'Riordan, T. (1971) *Perspectives on Resource Management*, London.
—— (1977) Resource Management, Unit 23, Open University Course D204, *Fundamentals of Human Geography*, Milton Keynes.
Orwell, G. (1937) *The Road to Wigan Pier*.
Penouil, M. (1969) 'An Appraisal of Regional Development policy in the Aquitaine Region', in Robinson, E. A. G. (ed.) *Backward Areas in Advanced Countries*, London.
Perloff, H. and Wingo, L. (1961) 'Natural Resource Endowment and Regional Economic Growth', in Spengler, J. J. (ed.) *Natural Resources and Economic Growth*, Washington, D.C.
Peters, W. C. (1978) *Exploration and Mining Geology*, New York.
President's Appalachian Regional Commission (1964) *Appalachia: a Report*, Washington, D.C.
Pounce, R., Upson, R., and Walker, C. (1976) 'Scottish Industry and North Sea Oil',

Trade and Industry, 414–19.
Richardson, H. W. (1965) 'Over-commitment in Britain before 1930', *Oxford Economic Papers*, 17, 237–62.
—— (1976) *Urban and Regional Economics*, Harmondsworth.
Robertson, J. G. (1974) Some Aspects of Mineral Production in Modern Britain, with particular reference to Yorkshire Potash and Merioneth Copper. Unpublished Ph.D. thesis, University of Hull.
Robinson, D. J. (1971) 'Venezuela and Colombia', in Blakemore, H. and Smith, C. T. (eds.) *Latin America: Geographical Perspectives*, London.
Sant, M. E. C. (1972) *Inter-regional Industrial Movement: the Case of the Non-Survivors*, Norwich.
Schumacher, E. F. (1973) *Small is Beautiful*, London.
Scottish Economic Bulletin (1977) 'The Impact of North Sea Oil-related Activity on Employment in Scotland', No. 11, 8–14.
Searle, G. (1975) 'Copper in Snowdonia National Park', in Smith, P. J. (ed.) *The politics of physical resources*, Harmondsworth.
Secretary of State for Energy (1978) *Energy Policy, a consultative document*, H.M.S.O., London.
Smith, D. M. (1971) *Industrial Location, an Economic Geographical Analysis*, New York.
Spengler, J. J. (ed.) (1961) *Natural Resources and Economic Growth*, Washington, D.C.
Spengler, J. J. (1967) 'Some Determinants of the Manpower Prospect, 1966–85', in Siegal I. H. (ed.) *Manpower Tomorrow: Prospects and Priorities*, New York.
Stohr, W. (1975) *Regional Development Experiences and Prospects in Latin America*, The Hague.
Thompson, W. (1968) 'Internal and External Factors in the Development of Urban Economies, in Perloff, H. S. and Wingo, L. (eds.) *Issues in Urban Economics*, Baltimore.
Tipton, F. B. (1976) *Regional Variations in the Economic Development of Germany during the Nineteenth Century*, Middletown, Connecticut.
Townroe, P. M. (1971) *Industrial Location Decisions: a Study in Management Behaviour*, Centre for Urban and Regional Studies, Birmingham.
Warren, K. (1973) *Mineral Resources*, Newton Abbot.
—— (1976) *The Geography of British Heavy Industry since 1900*, Oxford.
Wilson, M. G. A. (1968) 'Changing Patterns of Pit Location on the New South Wales Coalfields', *Annals of the Association of American Geographers*, 58, 78–90.
Wrigley, E. A. (1961) *Industrial Growth and Population Change*, Cambridge.
—— (1962) 'The Supply of Raw Materials in the Industrial Revolution, *Economic History Review'*, 15, 1–16.
—— (1965) 'Changes in the Philosophy of Geography', Chapter One in Chorley, R. J. and Haggett, P. (eds.) *Frontiers in Geographical Teaching*, London.
——(1969) *Population and History*, London.
Zimmerman, E. W. (1951) *World Resources and Industries*, New York.
Zuckerman Commission on Mining and the Environment (1972) *Report*.